639.2 B256

Barkin, J.

Saving glo   P9-CPY-026

reducing fishing capacity to

promote sustainability

WITHDRAWN

Saving Global Fisheries

Saving Global Fisheries
Reducing Fishing Capacity to Promote Sustainability

J. Samuel Barkin and Elizabeth R. DeSombre

CUYAHOGA COMMUNITY COLLEGE
EASTERN CAMPUS LIBRARY

The MIT Press
Cambridge, Massachusetts
London, England

© 2013 Massachusetts Institute of Technology
All rights reserved. No part of this book may be reproduced in any form by any electronic or mechanical means (including photocopying, recording, or information storage and retrieval) without permission in writing from the publisher.

MIT Press books may be purchased at special quantity discounts for business or sales promotional use. For information, please email special_sales@mitpress.mit.edu or write to Special Sales Department, The MIT Press, 55 Hayward Street, Cambridge, MA 02142.

This book was set in Sabon by the MIT Press. Printed on recycled paper and bound in the United States of America.

Library of Congress Cataloging-in-Publication Data

Barkin, J. Samuel, 1965–
Saving global fisheries : reducing fishing capacity to promote sustainability / J. Samuel Barkin and Elizabeth R. DeSombre.
  pages   cm
Includes bibliographical references and index.
ISBN 978-0-262-01864-7 (hardcover : alk. paper)
1. Sustainable fisheries. 2. Fishery management, International. 3. Overfishing.
I. DeSombre, Elizabeth R. II. Title.
SH329.S87B37   2013
639.2—dc23
                    2012026832

10 9 8 7 6 5 4 3 2 1

Contents

Acknowledgments

We are grateful to many people and institutions for their assistance in the writing of this book. Wellesley College provided funding for some of the research we undertook through the summer social science research program and the Educational Research and Development committee. Camilla Chandler Frost established the endowment that created the Wellesley College Environmental Studies Program and contributed to research opportunities for this book. And during most of the drafting stage, both of us were lucky to spend time in the Environmental Studies Program at Wellesley College, made up of an incredibly collegial group of people.

We appreciate feedback on early stages of this project from D. G. Webster, David Downie, International Studies Association panel attendees, and several anonymous reviewers for the MIT Press. And it is, as always, a delight to work with Clay Morgan, acquisitions editor extraordinaire at the MIT Press. Thanks as well go to our production editor, Deborah Cantor-Adams, and copyeditor, Annie Barva. We are also grateful to the Environmental Studies Section of the International Studies Association for the opportunity to present earlier versions of this work and to the broader global environmental politics community, which is a model for how collaborative and supportive academic research and teaching can be.

Jennifer Lee, Isabella Gambill, Ana Thayer, Shilpa Idnani, and Marsin Alshamary provided research assistance; Jessica Hunter was a formatting goddess and has been helpful to us in more ways than we can enumerate. Yuliya Rashchupkina compiled the index. Finally, we thank our favorite Cambridge area coffee shops—Voltage, Area 4, Flour, Crema Café, and High Rise—where much of this book was written and revised.

Acronyms

CCAMLR	Commission for the Conservation of Antarctic Marine Living Resources
CCSBT	Commission for the Conservation of Southern Bluefin Tuna
CPR	common-pool resource
EEZ	exclusive economic zone
EU	European Union
FAO	United Nations Food and Agriculture Organization
FOC	flag of convenience
GDP	gross domestic product
GEF	Global Environment Facility
IATTC	Inter-American Tropical Tuna Commission
ICCAT	International Commission for the Conservation of Atlantic Tunas
IOTC	Indian Ocean Tuna Commission
ITQ	individual transferable quota
IUU	illegal, unreported, and unregulated
MEY	maximum economic yield
MSY	maximum sustainable yield
NAFO	Northwest Atlantic Fisheries Organization
NEAFC	Northeast Atlantic Fisheries Commission
OECD	Organization for Economic Cooperation and Development
RFMO	regional fisheries management organization
SEAFO	Southeast Atlantic Fisheries Organization
TAC	total allowable catch
TURF	territorial use right for fishing

UNCLOS   United Nations Convention on the Law of the Sea
UNEP   United Nations Environment Programme
WCPFC   Western and Central Pacific Fisheries Commission
WTO   World Trade Organization
WWF   World Wildlife Fund

1

Introduction

The earth's oceans are overfished. Despite more than a half-century of extensive cooperation among the world's fishing nations, 80 percent of commercial fisheries are overexploited, significantly depleted, or fully exploited.[1] One particularly alarmist study predicted the complete collapse of global fisheries by 2048.[2] Other studies suggest that global populations of large predatory fish (those species such as tuna and swordfish that are subject to international management) are at a mere 10 percent of their preindustrial exploitation levels.[3] These individual species losses contribute to larger-scale losses within the ecosystem.

Although the most dire predictions of ecosystem collapse may (or may not) be overstating the problem, it is certainly the case that current rates of fishing are in the aggregate unsustainable. Although some specific fisheries have become increasingly well managed, international efforts have failed to reduce overexploitation at a global scale. Smaller successes have not scaled up, and fish species across many parts of the ocean have declined.

This dramatic depletion of ocean ecosystems has serious consequences for people. Fish are an important source of food for a growing human population, accounting for 15 percent of the global consumption of animal protein; in some places, fish can constitute fully half of the

---

1. FAO, *The State of World Fisheries and Aquaculture* (Rome: FAO, 2009).

2. Boris Worm, Edward B. Barbier, Nicola Beaumont, J. Emmett Duffy, Carl Folke, Benjamin S. Halpern, Jeremy B. C. Jackson, et al., "Impacts of Biodiversity Loss on Ocean Ecosystem Services," *Science* 314 (2006): 787–790.

3. Ranson A. Myers and Boris Worm, "Rapid Worldwide Depletion of Predatory Fish Communities," *Nature* 423 (2003): 280–283.

animal protein consumed by a state's population.[4] The authors of a major collective analysis of the effect of fisheries depletion on marine biodiversity evaluated the effects that this fishery devastation has on the ecosystem services provided (to humans) by the ocean and concluded that the loss of marine biodiversity "is increasingly impairing the ocean's capacity to provide food, maintain water quality, and recover from perturbations."[5]

Understanding the Problem

How did the global fisheries crisis become so bad? The situation has been characterized as "too many boats, chasing too few fish."[6] The United Nations Food and Agriculture Organization (FAO) in its 1999 International Plan of Action for the Management of Fishing Capacity called on states to develop National Plans of Action to manage fishing capacity,[7] but more than a decade later only three states have filed such plans with the FAO.[8] By some estimates, global fishing capacity is at 250 percent of the sustainable level.[9] There are too few fish because they have been overfished by centuries of commercial fishing efforts and subject to decades of unsuccessful management. There are too many fishers because ocean fish are unowned until they are captured, which leads to a global race to catch them before someone else does. Bigger, faster vessels can get to the fish first and capture them fastest. But once the fish are caught, these ships are left with nothing to do and are underutilized capital. So states subsidize their fishers to assist them in this endeavor, allowing bigger vessels and

4. FAO, *The State of World Fisheries and Aquaculture*, 58–65.

5. Worm et al., "Impacts of Biodiversity Loss on Ocean Ecosystem Services," 787.

6. By, among others, Joe Borg, EU Commissioner for Fisheries and Marine Affairs. See David Charter, "Too Many Boats Chasing Down Too Few Fish," *London Times*, 16 April 2009, http://www.timesonline.co.uk/tol/travel/news/article6101802.ece.

7. FAO, "International Plan of Action for the Management of Fishing Capacity," 1999, http://www.fao.org/fishery/ipoa-capacity/legal-text/en.

8. FAO, "National Plans of Action—International Plan of Action for the Management of Fishing Capacity," 2011, http://www.fao.org/fishery/ipoa-capacity/npoa/en; the three states are the United States, Namibia, and Indonesia.

9. WWF, *Hard Facts, Hidden Problems: A Review of Current Data on Fishing Subsidies* (Washington, DC: WWF, 2001), ii.

fancier technology that enable stocks to be depleted faster, leading the fishing race to become even more necessary.

This dynamic is exacerbated by two other causes of subsidization. Governments and development banks fund growth in fishing fleets as a form of economic development, causing growth in the number and effectiveness of vessels even though there are fewer and fewer fish for them to catch. Although the logic of using marine capture fisheries as a source of economic development ultimately fails, the lure of developing an industry to exploit a resource that seems readily available nonetheless often drives subsidies in the short term. And more broadly, fishing is often a cultural emblem and a vestige of a hunter-gatherer past. This romantic view of fishing and fishing communities is often able to generate subsidies from governments even when economic logic suggests that support for the industry is misplaced.

Existing patterns of international cooperation try to limit the number of fish that can be caught, while simultaneously encouraging—explicitly through subsidies or implicitly through policies that reward greater capitalization—increased fishing capacity. They may succeed at protecting individual species or regions, but at the cost of shifting existing capacity to new fish stocks or new areas. These solutions focus on the wrong part of the problem. Capacity is moved around rather than being effectively limited or reduced. We call this phenomenon a balloon problem because when fishing capacity is squeezed in one place, it balloons out somewhere else. If we are to save global fisheries, we need to change fundamentally the way we approach fishing at the domestic and international level. We need to see fisheries as long-term environmental resources embedded in ecosystems rather than as solutions to short-term domestic employment crises or the embodiment of noble cultural emblems. We need to stop subsidizing fishers and figure out how to get people out of the industry.

Our Proposed Solutions

In order to be effective, the international regulatory system, in which regional fisheries management organizations (RFMOs) attempt to reduce overexploitation of marine fisheries by limiting amounts that fishers can catch and the ways in which they can fish, must be reformed. It needs to be augmented by mechanisms to reduce fishing capacity and to decrease the number of people and amount of capital employed in the industry. Some successful efforts at national fisheries management, entirely within the jurisdiction of individual countries, already undertake these measures. But there are no equivalent mechanisms at the international level. Individual countries can only be effective at managing fisheries that are wholly within their waters. International fisheries, including fish that live all or part of their lives in the high seas as well as fish that swim across national marine boundaries, cannot be protected without global mechanisms to reduce fishing capacity.

We argue for a global fisheries organization as the institutional mechanism that can best address the problem of international overcapacity in the fishing industry. This organization would have three key aims. The first would simply be to advocate for less capacity—for a view of and regulatory setting for global fisheries that focuses on sustainability. The second would be to encourage and coordinate cooperatively reducing subsidies to the fishing industry, which may account for close to half of the industry's total value. Only a comprehensive approach to reducing subsidies can avoid the problem of free riders, countries that try for a competitive advantage by subsidizing their own industries while other countries do not. The third would be to work toward a global system of fishing permits, or individual transferable quotas (ITQs), for international fisheries. By more closely aligning fishers' interests with the demands of sustainable resource management, an ITQ system would help to remove capacity from the system while at the same time improving the fishing industry's financial as well as environmental sustainability.

We need to create a new set of mechanisms for managing practice within the industry that might have a chance of success. This new set of mechanisms must augment and complement the existing pattern of microregulation in which fishers' access to specific fish stocks is restricted, but without any concomitant decrease in the global supply and capacity of fishers. The current approach has been demonstrated both theoretically and empirically to be ineffective on its own and therefore insufficient. We need, in addition, a focus on the macroscale by thinking in terms of fishing capacity as well as of managing specific stocks. This focus involves both conceptualizing the problem differently and creating forms of regulation that can reduce the global supply of fishers and work to align their incentives with long-term management of global fisheries.

The proposal we make in the second half of this book, for a global fisheries organization, is ambitious. It calls for a fundamental restructuring of global fisheries politics and of the institutions that govern international fisheries management. It is not something that can be achieved easily and may well not be something that can be achieved either under current international political conditions or in one step. Nor do we propose here the details of institutional design or funding for a global fisheries organization or the specific mechanics of an agreement to reduce subsidies or create an international ITQ system, in part because such a proposal would at this point be premature. Rather, our goal here is to make the case that a macroapproach to international fisheries management is necessary and to suggest one way in which we might think about macroscale regulation. Our goal, in other words, is to promote a conversation, not a specific conclusion;[10] to provide some foundations rather than a complete architecture.

We are not proposing an all-or-nothing solution. Although the ideas for macroregulation and for a global fisheries organization would work best in concert, any progress made in cooperative efforts to reduce capacity would help improve international fisheries management. Getting the conversation going in a way that changes discourse and thinking to focus on capacity would be a good start and would constitute a partial success. Any cooperative efforts to limit, reduce, and ultimately eliminate subsidies, even in the absence of a global fisheries organization, would constitute a more significant success. Both conversation and cooperation would provide effective building blocks to construct discussions for an international ITQ or ITQ-like mechanism.

We must also stress that we are not suggesting that our approach should either replace existing microlevel fisheries regulation or displace alternative ideas for improving global fisheries management. Our proposal for international ITQs builds on regulation by RFMOs and needs effective microlevel management to work. And subsidy reduction alone

---

10. We are promoting this conversation rather than starting it. Previous discussions of macroscale regulation of international fisheries include G. T. Crothers and Lindie Nelson, "High Seas Fisheries Governance: A Framework for the Future?" *Marine Resource Economics* 21 (2007): 341–353; Robin Allen, James Joseph, and Dale Squires, *Conservation and Management of Transnational Tuna Fisheries* (Malden, MA: Wiley-Blackwell, 2010); and James Joseph, *Past Developments and Future Options for Managing Tuna Fishing Capacity, with Special Emphasis on Tuna Purse-Seine Fleets* (Rome: FAO, 2005).

will not necessarily help any given overexploited species. A system for capacity management can be usefully augmented by mechanisms such as marine protected areas, which allow stocks safe spaces in which to breed and maintain reserves of genetic and ecological diversity. Finally, social movements such as the sustainable-seafood movement can provide invaluable political momentum on which to build efforts to address overcapacity in global fisheries.

Most important, our goal is to make clear that doing a better job of what the current regulatory system is doing is not only insufficient, but can be counterproductive. Regional management of fisheries when fishing vessels are able to operate globally simply encourages vessels to chase available fish. Without decreasing capacity, international regulation of fish, even if undertaken globally, cannot succeed. If we are to save global fisheries, we need to change the way fisheries management is done.

Book Overview

Part I of the book diagnoses the problems that collectively underpin the current overcapacity in and underprotection of global fisheries. We begin in chapter 2 with an overview of current patterns of international regulation designed to manage global fisheries and the difficulties faced by international regulators. These difficulties include the implications of the common-pool resource (CPR) structure of global fisheries—the mismatch between individual and collective (and short-run and long-run) incentives that are at the core of many international environmental problems. These problems are compounded by unusually prevalent problems of scientific uncertainty and major hurdles to the monitoring and enforcement of rules. International fisheries management is dominated by RFMOs, which focus on the management of either region-specific or species-specific individual fisheries. These organizations rely on scientific advice about sustainable catch levels, create rules about things such as quotas and fishing methods, and attempt to monitor and enforce fishing behavior and rules. But even when this pattern of management works at the microscale by reducing the pressure on specific species, it fails at the macroscale issue of reducing the pressure on global fisheries resources.

Chapter 3 evaluates this broader point about the ineffectiveness of microregulation by examining the political economy of fisheries regulation.

A discussion of the economic implications of CPRs and the challenges faced by environmental economics in dealing with CPR problems sets the broad outline of the problem. The global nature of ocean fish resources magnifies these problems by embedding them in a multilevel interaction in which states face both domestic constituencies and international obligations; the domestic politics of fisheries often leaves fishers with barriers to exit from the industry. We ultimately argue that the type of microregulation practiced by RFMOs and states creates what we call a balloon problem. International regulation focuses on reducing fishing effort with respect to specific species, but this sort of regulation creates a balloon effect: if fishing effort is squeezed in one place, but fishing capacity remains constant, fishers, rather than exiting the industry, will look for somewhere else to fish. Reducing fishing pressure on particular fish stocks, in other words, has the effect of transferring the pressure to other stocks rather than reducing overall global levels of overfishing.

Why have so many national fisheries regulators not responded to clear evidence of overfishing by acting to reduce fishing capacity, either individually or collectively? Chapter 4 argues that a key reason for this lapse is that most fisheries regulators focus on fisheries as an industry to be protected and developed rather than on fish as a resource to be conserved. In short, there is a widespread problem of industry capture of both national and international fisheries regulators. Fisheries regulators are often as concerned about supporting and mollifying their constituents as they are about long-term management of natural resources. Fishers have a political voice, but fish do not. At the domestic level, subsidies are the primary effect of this political capture. At the international level, regulations within RFMOs are much laxer than scientific advice recommends. In both cases, fishing pressure increases and stocks decline.

Chapter 5 discusses an additional reason that it has been so difficult for rational actors to seek restraint in fishing: the culture of fishing and of fishing regulation. Parts of this issue are structural. Collective-action problems are exacerbated by the CPR nature of fisheries, and domestic fishing interests are able to garner much more political power than their size or economic importance would suggest. Other parts of the problem are cultural. Fishing is often culturally embedded as a pursuit that must be supported for social and symbolic reasons, despite the economic and environmental reasons to limit it. The discourse of fisheries management also contributes to the problem, with academic approaches to analysis of fisheries problems focusing too much on the natural science of fish and not enough on the social science of fishers.

These political, economic, and cultural factors contribute to regulation for the good of the fishing industry rather than for the common good; one major effect of this approach is the subsidization of all aspects of the fishing industry, addressed in chapter 6. Many governments actively subsidize their fishing industries, creating capacity far in excess of what the market would support on its own, let alone what the resource base can support. This subsidy is sometimes overt, through direct subsidies for vessels or fishing activities. It is often indirect, though. Governments fund the modernization of fleets, which inevitably increases capacity even if it reduces the number of vessels at sea. They subsidize their fleets by buying access to foreign waters or providing other services to the industry that are paid for through general taxation. And they keep fishing communities afloat well beyond the time when they are no longer commercially viable, through social welfare models designed to protect existing industries rather than develop new ones. These subsidies create a vicious cycle in which those who receive them are able to increase their political voice for continuing subsidies and for pursuing short-term fishing policies that deplete fisheries and thus lead to demands for increase subsidization. Decreasing subsidies will be extremely difficult, but it is central to the ability to decrease capacity and thus manage fisheries for the long term.

One problematic form of subsidization is the use of fisheries as a tool for economic development, discussed in chapter 7. It is understandable that governments and aid organizations might look to fishing expansion when seeking development. Expanding fishing capacity yields employment benefits immediately, whereas the costs of excess capacity are felt over a longer period. We argue, however, that pursuing fishing as development policy is misguided. Unlike for most other industries, the fixed limit on the size of fish stocks means that fisheries have a built-in limit to growth. As soon as a fishing industry is successfully developed, therefore, it must immediately be reined in, or it will begin to overexploit its resource. This dynamic makes fisheries development unsuitable as an engine of continuing economic growth. Arguments that developing countries should be allowed the same opportunities to develop as industrialized countries miss the broader point that using fisheries as development policy is not only problematic for the health of global fish stocks, but also counterproductive to the development opportunities it is supposed to create.

Part II of the book develops our proposals for saving global fisheries by addressing these problems. Chapter 8 argues that we need to embed the system of regional fisheries management within an overarching global institution. Looking at fisheries management as a global issue entails embedding the current system of RFMOs and national regulation within a global rather than regional multilateral framework. The new global fisheries organization we propose would need to be oriented to the environment rather than to industry. Among its other tasks (such as information gathering and enforcement), a new institution would help coordinate international binding efforts to reduce subsidies and manage capacity.

We then make an extended argument for the advantages of creating property rights as the way for this global organization to functionally manage capacity and regulate catch levels globally. Chapter 9 examines the arguments for privatization as one type of solution to managing CPR problems and how such an approach might function usefully in a fisheries context. Several decades of experience with domestic-level ITQ systems for managing fisheries suggest the advantages of such programs for aligning fishers' interests with long-term fishery conservation. This chapter evaluates design options for such programs and their successes (and difficulties) at the domestic level.

Chapter 10 then discusses how a global organization might realistically be created. Scaling ITQ systems up to the international level presents a number of technical and political challenges. This chapter discusses parameters for how this organization might be designed. Such a system would allow RFMO regulation to work effectively by deflating the balloon problem. Quotas on some species would not create undue pressure elsewhere in the oceans because overall capacity would be limited accordingly. The new system would also decrease pressure for subsidization (which in turn would decrease pressure for unsustainable catch levels) and would create opportunities for fishers to exit the system in an economically beneficial way.

We are under no illusion that the approaches we propose will be easy, but they are necessary. Chapter 11 recapitulates the key elements in this book and points out both the potential of and some of the limits to our arguments. Current forms of international management of fisheries are unsustainable both environmentally and economically; they harm the human populations they are supposed to help and fail to create long-term economic development or sustenance. If we hope to save global fisheries, we need to address the pathologies of the current domestic and international approaches to ocean fisheries and undertake new efforts to protect ecosystems and communities. This book is our proposal for how these goals might be undertaken.

I

The Problems

2

International Regulation

The regulation of international fisheries faces the same difficulty as all international regulation: the absence of a central authority able to make and enforce binding rules. Any regulation must be by collective action among states. But even beyond this general problem, there are two reasons that fisheries are particularly difficult to regulate. The first is that they are CPRs, meaning that access to fisheries is difficult to restrict, which makes depletion likely. Although states benefit collectively in the long run from cooperative action to protect a fishery, individual fishers and individual fishing states often gain the most by exploiting the resource in the short run, especially if they are concerned that collective management will not work. Or they may choose to remain apart from cooperative efforts undertaken by others and thereby free ride. Cooperative outcomes are therefore expected to be underprovided and noncompliance prevalent. The second difficulty of fisheries regulation is that fisheries involve a high degree of uncertainty. This uncertainty can be about the health (and general characteristics) of the resource, about actual fishing behavior, and even about the relationship between regulatory measures and environmental outcomes—all of which combine to make already difficult resource management even more difficult.

The primary international mechanisms in place to deal with these difficulties and regulate international fisheries are called RFMOs. These organizations generally include scientific bodies to deal with the uncertainties of fisheries management and regulatory bodies to create rules to prevent or minimize abuse of the commons. But their regulatory structures and processes can actually add to the difficulties of regulation. RFMOs often set total allowable catches (TACs) in fisheries that exceed the catch amounts that scientific analysis (often conducted within the organization itself) has indicated are sustainable, as documented and discussed in chapter 4. Even those too-high fishing levels are exceeded through noncompliance within the regulatory process or because fishers operate outside that process. More important, RFMOs focus on what we call microregulation, the regulation of what fishers do with specific species in specific regions, a focus that is structurally flawed.

This RFMO approach to regional (and species) regulation ultimately contributes to the broader problems, discussed further in chapter 3, of fishers changing regions or species to continue their fishing effort when they face regulation pertaining to one area or species. Chapters 3 and 4 examine how international fisheries regulation came to focus on regions and species and fishing behavior within them (what we call the microscale) and what the problems with an exclusive focus on this scale are. This chapter takes the preliminary step of outlining the problems fisheries face because of CPR issue structure and uncertainty, and the institutional structure of the RFMO system as a response to these problems.

Issue Structure

A CPR is defined by two characteristics: it is subtractable (also called "rival"), and it is nonexcludable. A resource is considered subtractable when its use by one actor diminishes other actors' ability to use the resource. Subtractability is thus inherent in the resource itself and how we use it. On the oceans, a fish caught by one fishing vessel is no longer available for capture by others, nor does it remain in the water to reproduce and thereby create more fish for everyone.

A resource is considered excludable if an owner or possessor can keep other potential users from access to it. There are two components to excludability, the legal and the practical. International fisheries are non-excludable (potential users cannot be prevented from access) on both counts. In practical terms, excluding fishing vessels from access to fish is difficult, particularly when the vessels are far from shore, when regulators have insufficient resources, or when the fish swim across jurisdictions. The ocean is vast, and it is difficult to determine which vessels are fishing for which species at any point in time. So even in the context of a regulatory regime that might grant the legal authority to exclude vessels from access to fish, actually doing so is at best complicated. Fisheries organizations create lists of excluded vessels but rarely, if ever, try to do anything to remove them if they are fishing in the area from which they have been excluded (although states often do so within their nationally controlled waters). The closest RFMOs have come to any kind of exclusion is through restricting trade in fish products from ships registered in nonmember states.[1]

The high seas were unowned historically; no states or fishers had individual rights to the resources there, so they did not have the right to keep others out. Legal structures can change over time, however, affecting whether a resource is legally excludable or not. The high seas are currently still unowned, but jurisdiction—and therefore the right to exclude—was established practically over two-hundred-mile exclusive economic zones (EEZs) in the 1970s. This effective jurisdiction was then made legal in 1982, when the United Nations Convention on the Law of the Sea (UNCLOS) formalized the creation of EEZs, granting states the legal authority to control access to the marine resources within two hundred nautical miles of their coastlines.[2] This measure put a substantial portion of the ocean and its resources under state control, but a vast swath of the open ocean remains unowned, and the high-value, highly migratory species live in or pass through that space or cross EEZ boundaries. Any rules created, such as through RFMOs, in the open ocean or across marine boundaries need to be created voluntarily by the states that decide to take them on.

The CPR nature of resources has many implications. Cooperation to address these types of issues is particularly difficult. With no overarching authority to mandate action on international issues, states need to agree to cooperate. A CPR's lack of excludability means that any actor that does not contribute to a cooperative outcome cannot practically be excluded from access to the resource, and the subtractability means that those noncooperators can diminish the resource. In other words, if most, but not all, states in the world get together and agree to restrict their catches of fish to levels at which the species can be sustainably caught, the resource can still be depleted by the few actors that do not restrict their

---

1. Elizabeth R. DeSombre, "Fishing under Flags of Convenience: Using Market Power to Increase Participation in International Regulation," *Global Environmental Politics* 5 (4) (2005): 73–94.

2. UNCLOS (1982), Article 57.

catches. The knowledge that nonparticipators can undermine cooperation can make it difficult to persuade those who have agreed to cooperate to sacrifice for species' long-term health; they may be rightly suspicious about the value of undertaking that sacrifice because they know their efforts may be undermined by others who do not participate. This aspect contributes to states' reluctance, even within the context of RFMOs, to agree to catch limits restrictive enough to protect fish stocks. Although everyone benefits in the long run from successful conservation, no one wants to be the only one to undertake action.

Thus, although collective-resource or environmental issues are sometimes incorrectly referred to as public goods,[3] CPRs are different from public goods. Unlike CPRs, public goods are not subtractable, which makes cooperation by a sufficiently motivated small group possible.[4] Even if others will free ride on the provision of the public good (precisely because they cannot be excluded from its benefits even if they do not participate in its provision), their lack of participation will not prevent the good from being provided in the same way that can happen with a CPR. It matters, therefore, to characterize fisheries (as well as most environmental issues[5]) as CPRs.

Noncompliance with regulation has the same effect on the resource as nonparticipation in the context of a CPR. Any actor taking more than a sustainable percentage of the fish stocks contributes to decreasing the overall availability of fish. And the mere possibility that someone might not be participating or complying with collective-management measures is enough to cause a potential problem. If it is difficult to determine who

---

3. See, for example, Joseph Stiglitz, "A Neoclassical Analysis of the Economics of Natural Resources," in *Scarcity and Growth Reconsidered,* ed. V. K. Smith, 36–66 (Baltimore: Johns Hopkins University Press, 1979), and Scott Barrett, *Environment and Statecraft: The Strategy of Environmental Treaty-Making* (Oxford: Oxford University Press, 2003). Others refer to fisheries as "impure public goods": United Nations Industrial Development Organization, *Public Goods for Economic Development* (Vienna: United Nations Industrial Development Organization, 2008), table 1.1.

4. Russell Hardin, *Collective Action* (Baltimore: Johns Hopkins University Press, 1982); Robert O. Keohane, *After Hegemony: Cooperation and Discord in the World Political Economy* (Princeton, NJ: Princeton University Press, 1984).

5. J. Samuel Barkin and George E. Shambaugh, eds., *Anarchy and the Environment: The International Relations of Common Pool Resources* (Albany: State University of New York Press, 1999).

has lived up to their obligations, it can be difficult to persuade actors to take obligations on in the first place or to live up to them once they have.

Garrett Hardin's seminal work on the difficulty of cooperation in managing CPRs under conditions of anarchy frames the issue as a "tragedy of the commons," noting that "freedom in the commons brings ruin to all."[6] Hardin's and others' dire predictions for cooperation over CPRs should not be overstated; cooperation to protect shared depletable resources has certainly been possible in some circumstances, but it is clear that these types of collective goods have been underprovided. In the case of fisheries, the CPR structure can help explain why.

The CPR nature of global fisheries has additional implications. Compared to collective goods with other structures, CPRs increase the importance of concern among actors for relative gains, shorten the shadow of the future, and increase the likelihood of conflict within the resource.[7]

Because CPRs can be depleted, the time that it takes to come up with a management process matters. This issue structure also gives differential negotiating power to actors depending on their interest in creating a solution to the problem. Those who do not depend as heavily on the resource can hold out in a contentious negotiation because they know that the longer they wait, the more depleted the resource will become. Others who expect to rely more heavily in the long term on that resource may be more likely to agree to the bargaining position of the actor who can indicate indifference. A study of fishing conflicts found that states whose fishers had the capacity to fish elsewhere (and therefore did not need the fishery resources about which there was a management conflict) were better able to achieve their preferred outcome in negotiations over management.[8]

Those credibly able to commit to remaining outside of collective processes have similar bargaining power specific to negotiations to address CPRs, precisely because their ability to remain outside of cooperative arrangements can doom the resource despite others' efforts to conserve it.

6. Garett Hardin, "The Tragedy of the Commons," *Science* 162 (1968), 1244.

7. J. Samuel Barkin and George E. Shambaugh, "Hypotheses on the International Politics of Common Pool Resources," in *Anarchy and the Environment*, ed. Barkin and Shambaugh, 11.

8. J. Samuel Barkin and Elizabeth R. DeSombre, "Unilateralism and Multilateralism in International Fisheries Management," *Global Governance* 6 (2000): 339–360.

For the same reason, benevolent hegemonic efforts to protect CPRs may not work in the same way that they can with the provision of public goods.[9]

Collective Long-Run Incentives

Part of the reason for the tragedy of the commons, as identified by Hardin, is the individual versus collective incentive structure. This process is akin to the game theoretic structure of the prisoner's dilemma.[10] Because fish are a renewable resource, if a sufficient number of a given species remains to reproduce, the fish stock will continue such that fishing can be done indefinitely. For that reason, it is in the interest of fishers who want to continue to fish forever to keep their levels of fishing to that sustainable level. But a number of collective problems can interfere with that incentive.

The first is that individual fishers want to avoid what in game theory is called "the sucker's payoff":[11] a situation in which they reduce their catch in order to protect the resource, but others do not. Undertaking personal sacrifice in the short run to protect the resource and ensure its long-run survival (and therefore your ability to use it) is one thing; undertaking personal sacrifice in the short run and failing to achieve resource protection in the long run is a foolish decision. Given the subtractability of global fisheries, any nonparticipants can undermine the ability to protect a fish stock. Nonexcludability means that it is difficult to ensure full participation or compliance with whatever cooperative measures are taken up.

At the same time that actors realize that it is foolish to cooperate when others are not, they also realize that the best outcome they can experience would be for everyone else to protect the fishery while they stay outside of the cooperative arrangements. The overall health of the fish stocks would improve because most fishers are practicing restraint, and

---

9. Barkin and Shambaugh, "Hypotheses on the International Politics of Common Pool Resources," 14–16.

10. Russell Hardin, *Collective Action* (Baltimore: Johns Hopkins University Press for Resources for the Future, 1982).

11. Charles Lipson, "International Cooperation in Economic and Security Affairs," *World Politics* 37 (1) (1984): 1–23.

so a free rider would get a double benefit: a protected stock and unfettered access to it. But if everyone simultaneously seeks this best outcome and attempts to avoid the sucker's payoff, little cooperation is likely. This standard prisoner's dilemma set of preferences[12] is expected to lead to an outcome in which cooperation is underprovided, even from the relevant actors' perspective.

The collective nature of states makes this cooperative structure more difficult as well. Even if the state as a whole would benefit collectively from protection of a fishery over the long term, individual fishers will suffer in the short term. Those fishers represent a concentrated interest politically: they care more about this issue than any other and will vote and organize and protest in order to achieve the goal they desire. States are receptive to these protests because these groups can represent important voting blocks. States with political representation by geographic area (such as the US Congress, which elects senators by state and representatives by districts) produce political decision makers who are especially attuned to the needs of the primary industry in their districts or states.[13] In addition to the difficulty that this problem creates for states in choosing how to act in international negotiation, states must face not only their domestic audiences, but several different international ones, as explained in greater detail in chapter 3.

Uncertainty

Uncertainty creates problems for international environmental regulation generally, and these problems magnify the difficulties in cooperation discussed earlier. Fisheries are subject to a particularly wide variety of factors about which there is uncertainty, including estimates of a fishery's health and of the abundance and characteristics of fish, predictions about stock abundance over time, and information about fishers' behavior.

The complexity of understanding basic information about a given fishery should not be understated. Population estimates for species are

---

12. Robert Axelrod, *The Evolution of Cooperation* (New York: Basic Books, 1984).

13. Andrew Moravcsik, "Why Is U.S. Human Rights Policy so Unilateralist?" in *Multilateralism and U.S. Foreign Policy: Ambivalent Engagement*, ed. Shepard Forman and Patrick Stewart, 435–476 (Boulder, CO: Lynne Rienner, 2002).

difficult under the best of circumstances, even when the species live where they can be seen by those doing the counting. Fish live in the murky depths of an opaque ocean and can move long distances, sometimes at high speeds. Simply assessing the number of fish in a population can be difficult.

As important as the number of fish at any given time is their life cycle. How many young do they produce (fecundity), and what proportion survives? How many years does it take for individuals to reach reproductive maturity? The orange roughy fell victim to poor knowledge on this measure. Availability of orange roughy on the market increased dramatically as deep-sea trawling harvested previously inaccessible fish species, and the species became popular on restaurant menus. But its fecundity is low, and the species reaches reproductive maturity only between twenty and forty-four years of age.[14] If fish are caught before they have the chance to reproduce, sustainability is impossible.

How those population estimates change over time and why are subject to even more uncertainty. One of the primary sources of information about the abundance of fish is those who are catching them: Is it easier or harder to catch fish this year? The easier they are to catch, the more abundant they must be. That measurement is less reliable to use when fishing technology changes rapidly. More fish may be caught in a shorter period of time not because the stock is more abundant, but because of the use of technical assistance (spotter airplanes, sonar) and fishing gear (trawl nets, longlines) that makes the catch easier. Efforts are made to standardize the measurement as a "catch per unit effort," but it can be difficult, especially in times of major change in process, to determine how to measure "effort" across different technologies or seasons.

There are also circumstances in which fishers might not be the most reliable sources of information, especially information that might lead to increasingly strict regulation of their activities. When data about fishing behavior determine the closing date of the season, fishers may think twice before reporting data that will be sure to close off their fishing opportunities for the year. For example, in the case of early commercial whaling, a global catch quota was not divided into state quotas, and information

---

14. Trevor A. Branch, "A Review of Orange Roughy *Hoplostethus atlanticus* Fisheries, Estimation Methods, Biology, and Stock Structure," *African Journal of Marine Science* 23 (2001): 181–203.

from whalers about their catches determined when the whaling season would end.[15] So giving accurate information about whale catches would lead to decreasing opportunities to catch more whales. Information about catches can likewise impact the taxes that fishing vessels pay or, if it indicates catches beyond those allowed, may result in other penalties.[16]

A related problem is noncompliance. If scientific estimates of the effectiveness of fishing restrictions depend on estimates of abundance that combine knowledge of existing catches with population estimates, both of these pieces of information need to be complete. Noncompliance with reporting provisions is legion within international environmental organizations in general,[17] and fishery organizations are no exception. Worse, states or fishing vessels may report inaccurate information. A classic example comes from early whaling regulation.[18] In the 1965 whaling season, the minimum catch size for baleen whales was thirty-eight feet. That year 90 percent of the baleen whales caught were reported to be between thirty-eight and thirty-nine feet long, a statistical impossibility.[19] Whalers were clearly catching undersize (and therefore underage) whales and reporting them as just meeting the required minimum size. This situation posed problems not only for the stock itself (because individuals were taken before they had a chance to reproduce), but also for scientific studies of the stock because it was no longer clear at what size whales were actually being caught.

Several recent studies examine extreme underreporting of catches in the Japanese southern bluefin tuna longline fishery. The Australian Scientific Committee to the Commission for the Conservation of Southern Bluefin Tuna (CCSBT), for instance, concluded from an analysis of

---

15. International Convention for the Regulation of Whaling (1946).

16. Colin W. Clark, *The Worldwide Crisis in Fisheries: Economic Models and Human Behavior* (Cambridge: Cambridge University Press, 2006); Carl J. Walters and Steven J. D. Martell, *Fisheries Ecology and Management* (Princeton, NJ: Princeton University Press, 2004).

17. US General Accounting Office, *International Environment: International Agreements Are Not Well-Monitored*, GAO-RCED-92-43 (Washington, DC: US General Accounting Office, 1992).

18. Although whales are not fish, they operate as a fishery.

19. Patricia Birnie, *International Regulation of Whaling: From Conservation of Whaling to Conservation of Whales and Regulation of Whale Watching*, Vols. 1 and 2 (New York: Oceana, 1985), 338.

southern bluefin tuna sold at the Japanese wholesale fish markets that the number sold far exceeded the number reported as caught by Japan and the other states that provide tuna to those markets.[20] This overcatch was at least 100 percent.[21] A subsequent review undertaken by the Australian and Japanese governments concluded that significant overcatching and misreporting had taken place since the early 1990s.[22] There are important implications of this overcatching for scientific assessment of these stocks and others. Not only is there uncertainty about the extent of current and historical catches of this species, but it is possible that catches of southern bluefin tuna are misreported as catches of other species of tuna, thereby throwing off catch data for other fisheries.[23]

Far worse was a case of blatant whaling noncompliance that was exposed after the fall of the Soviet Union. Post-Soviet Russian scientists acknowledged that official state-level catch statistics were intentionally incorrect during the Soviet era so that Soviet whalers could cover up noncompliance. Among other transgressions, Soviet whalers caught twice the number of whales they actually reported and had a systematic process of misreporting the whale species they caught. In this time period, they caught whale species that were subject to a complete moratorium, such as humpback, right, and blue whales.[24] This misreporting made scientific estimate of stock recovery or future sustainable harvesting unreliable.[25]

The relationship between restriction of fishing and protection of a fishery is also uncertain: a fishery's decline may have a number of causes

---

20. CCSBT, *Comparison of CCSBT Catch Data with Japanese Auction Sales of Frozen SBT*, Working Paper no. CCSBT-EC-0510/25 (Deakin, Australia: CCSBT, 2005).

21. B. Jeffries, *Catches by Other Countries*, Australian Fisheries Management Organization (AFMA) SBT-FAG/2000/20 (Canberra: AFMA, 2000).

22. Tom Polachek and Campbell Davies, *Consideration of the Implications of Large Unreported Catches of Southern Bluefin Tuna for Assessments of Tropical Tuna, and the Need for Independent Verification of Catch and Effort Statistics*, Commonwealth Scientific and Industrial Research Organization (CSIRO) Marine and Atmospheric Research Paper no. 23 (Clayton, Australia: CSIRO, 2008).

23. Ibid., 11.

24. Paul Brown, "Soviet Union Illegally Killed Great Whales," *The Guardian*, 12 February 1994; authors' interview with Ray Gambell, International Whaling Commission secretary, Histon, UK, August 1993.

25. "Call Me Smiley," *New York Times Magazine*, 13 March 1994.

outside of fishing for that particular fish species. Fish can be caught as bycatch in another fishery (fish caught unintentionally when another species is being targeted).[26] Some fish populations can decline when the fish they eat have been overcaught or have disappeared for some other reason. Pollution can affect fish survival,[27] as can habitat loss, and climate change is likely to have a major impact on fishery health.[28]

Uncertainty thus makes regulation difficult in many ways. If regulators do not know basic information about the biology of the fish species in question, they may allow for catches that do not account for the characteristics of reproductive success. Stock estimates are an inexact science at best. If they are either unclear or incorrect (sometimes because of the incentive for fishers to obscure information or because the correct information is not collected for all ways in which fish are caught or might otherwise decline), even rules carefully set and perfectly complied with will nevertheless fail to protect the stock.

The Process of Regulation

The historical ruling principle of ocean management was the "freedom of the seas."[29] CPRs' structural characteristics were for the most part not a problem for high-seas fisheries before the middle of the nineteenth century because existing technology was not capable of depleting major fish stocks. Beginning at the end of the nineteenth century, however, states realized the necessity to cooperate about a variety of issues pertaining to the management of the oceans, such as shipping regulation, pollution prevention, and the regulation of national resources. The seas remain free unless

26. National Oceanic and Atmospheric Administration, US Department of Commerce, *Evaluating Bycatch: A National Approach to Standardized Bycatch Monitoring Programs*, National Oceanic and Atmospheric Administration (NOAA) Technical Memorandum NMFS-F/SPO-66 (Washington, DC: NOAA, October 2004).

27. M. Shahidul Islam and Masaru Tanaka, "Impacts of Pollution on Coastal and Marine Ecosystems Including Coastal and Marine Fisheries and Approach for Management: A Review and Synthesis," *Marine Pollution Bulletin* 48 (2004): 624–649.

28. FAO, *The State of World Fisheries and Aquaculture* (Rome: FAO, 2009).

29. Hugo Grotius, *Mare Liberum* (1618; reprint, Oxford: Oxford University Press 1916).

they are the subject of negotiated agreement among states to restrict the activity of vessels or other activity in specific ways.

The foundational international agreement upon which current international fisheries management is based is UNCLOS. As well as allowing for two-hundred-mile EEZs, UNCLOS also codifies the legal space for fisheries to be managed by RFMOs. But these organizations' legal and practical authority to control access to the fisheries in question (i.e., to make fisheries resources excludable) is limited. Although UNCLOS mandates that states "cooperate with each other in the conservation and management of living resources in the areas of the high seas,"[30] there is disagreement on the extent of that legal obligation (especially for states that have not ratified UNCLOS).[31] Moreover, the obligation to cooperate does not necessarily create the legal authority to keep fishing vessels out of areas where their states have not accepted the obligations established by the relevant RFMOs.

The associated United Nations Agreement for the Implementation of the Provisions of the United Nations Convention on the Law of the Sea of 10 December 1982 Relating to the Conservation and Management of Straddling Fish Stocks and Highly Migratory Fish Stocks (often referred to either as the Straddling Stocks Agreement or as the Fish Stocks Agreement) elaborates on this obligation, especially with regard to species that migrate through EEZs or travel long distances in the ocean.[32] This agreement reinforces RFMOs' primacy for managing international fisheries and states' duty to cooperate with these organizations,[33] but it does not create practical mechanisms for enforcing that obligation. States that are not members of a given RFMO are likewise instructed to prohibit their vessels from fishing in its regulatory area,[34] but there is no particular consequence for not doing so. Moreover, many states have not ratified the Straddling Stocks Agreement (some have not even ratified UNCLOS),

---

30. UNCLOS (1982), Article 118.

31. Rosemary Rayfuse, *Regional Allocation Issues, or Zen and the Art of Pie-Cutting*, Faculty of Law Research Series, Paper no. 10 (Sydney: University of New South Wales, 2007), 3.

32. United Nations Convention on the Law of the Sea of 10 December 1982 Relating to the Conservation and Management of Straddling Fish Stocks and Highly Migratory Fish Stocks (1995) (hereafter "Straddling Stocks Agreement").

33. Ibid., Article 8.

34. Ibid., Article 17.

and so the extent to which these provisions apply to nonratifying states is questionable.

Almost all international regulation of fishing is currently done by RFMOs. These organizations regulate based on species or region or a combination of the two. Some (such as the Northwest Atlantic Fisheries Organization [NAFO]) have regulatory authority only over fish in international waters, whereas others (such as the International Commission for the Conservation of Atlantic Tunas [ICCAT]) regulate specific species in both international and national waters. RFMOs cover most of the world's oceans and commercially important international and migratory fish stocks. Depending on how you count, approximately nineteen RFMOs have stock management (rather than just scientific research) as a mandate.[35]

There are many similarities across RFMOs in how they are structured and how they regulate. Almost all RFMOs have open membership—any state with vessels fishing in the region can join. Even the Commission for the Conservation of Antarctic Marine Living Resource (CCAMLR), which technically restricts membership to those with an interest in the region,[36] is generally open to membership requests and more concerned about ensuring it includes all states engaged in fishing in the region than excluding states that are not. Most RFMOs have two components that are important in this context: a scientific committee and a commission. The scientific committee is designed to address some of the uncertainties that make fisheries regulation difficult. The commission is designed to address the CPR issue by creating limits on the withdrawal of fish from the commons.

Setting Catch Limits

Most RFMOs have scientific committees that are charged with either conducting or, more frequently, aggregating the relevant scientific research on stock health to determine what the maximum sustainable yield (MSY) of fish that can be taken from the fishery is for a given year. They make recommendations to the commission about what that amount is. (Some problems with the concept of MSY are discussed in chapter 3.)

35. FAO, *The State of World Fisheries and Aquaculture* (Rome: FAO, 2008).

36. Convention for the Conservation of Antarctic Marine Living Resources (1980), Article XXIX.

The commission passes any actual regulations. RFMOs set out in their founding documents what types of rules they can adopt, but these rules most frequently involve catch limits, opened or closed areas or seasons for fishing, size limitations on the fish caught, gear restrictions, and (less frequently) bycatch limits.

Although the process and rules vary across RFMOs, in general each member state has one vote in the commission. The commission usually meets annually (or at least every two or three years), and regulations are passed at each meeting. This process gives the commission the ability to adapt to new information about improving or declining stock health as it creates regulations and is more responsive than a process that produces regulations that last for a longer period.

Although a trend in recently created RFMOs is to require unanimity to pass measures, many RFMOs adopt fishing regulations by less than unanimous voting. These RFMOs almost always include an objections procedure (also called a reservations procedure) by which states that did not vote in favor of a rule that nevertheless passed can opt out of being bound by it. The process in ICCAT is illustrative. A two-thirds majority vote by commission members is required to pass regulations.[37] A regulation takes effect six months after passage, but during that time any state can lodge an objection, which means that the state chooses not to be bound by that obligation. Because this form of regulatory free riding would disadvantage those who continue to abide by a regulation to which another state has objected, the period of time before a regulation becomes binding is increased by an additional sixty days, during which time other states may also choose to object; after that time period, the obligations bind all states that have not chosen to object.[38] If a majority of states objects to a provision, it is not binding on any states.[39]

Other RFMOs with objections procedures include NAFO,[40] the Northeast Atlantic Fisheries Commission (NEAFC),[41] and the Indian Ocean

---

37. International Convention for the Conservation of Atlantic Tunas (1966), Article VIII(1)(b)(i). These regulations are called "recommendations," but they are legally binding on member states.

38. Ibid., Article VIII(2) and (3).

39. Ibid., Article VIII(3)(g).

40. Convention on the Future Multilateral Cooperation in the Northwest Atlantic Fisheries (1978), Article XII.

41. Ibid. (1980), Article 12.

Tuna Commission (IOTC).[42] This objections process is controversial;[43] it can lead to fishing seasons in which some of the major fishing states are not bound by the governing regulations in an area. But these provisions exist because it is unlikely that states would agree in advance to be bound by rules created by an international organization in which they can be outvoted. Nonunanimous voting can be useful as a decision-making rule because one recalcitrant state cannot prevent a regulation from going forward. The objections procedure is a safety valve that lets states know that if they really need to, they can avoid being bound by a specific regulation; they may then be more willing to agree to regulation in general.

There are a number of stages at which rules adopted by RFMOs are insufficient to address the conservation of resources under their purview. The first stage comes in the decision process the commission makes about the TAC in response to the recommendation from the scientific commission. Commission TACs are almost always higher than recommended by scientific advice.[44] For example, in 2004 the CCSBT's Scientific Commission expressed concern about low levels of fish reaching the age and size where they were available for commercial harvest for several years running but could not be certain of the likelihood of stock decline if current catch levels persisted. The commission noted that such a decrease "would have significant impacts on the fisheries and the communities that relied on it. Any decision to reduce the TAC as a management response could not, therefore, be taken lightly."[45] Chapter 4 further documents this phenomenon and elaborates some of the structural reasons for it.

Even apart from this problem, the CPR and short-run versus long-run incentive structures both decrease the odds that commission members will agree to regulations that are as strict as would be required to successfully

---

42. Agreement for the Establishment of the Indian Ocean Tuna Commission (1991), Article IX.

43. Howard S. Schiffman, *Marine Conservation Agreements: The Law and Policy of Reservations and Vetoes* (Leiden: Nijhoff, 2008).

44. FAO, Marine Resources Service and Fisheries Department, *Review of the State of World Fishery Resources: Marine Fisheries*, FAO Fisheries Circular no. 920, FIRM/C920 (Rome: FAO, 1997); Robert Alps, Laurence T. Kell, Hans Lassen, and Innar Liiv, "Negotiation Framework for Baltic Fisheries Management: Striking the Balance of Interest," *ICES Journal of Marine Science* 64 (2007): 858–861.

45. A. Willock and M. Lack, *Follow the Leader: Learning from Experience and Best Practice in Regional Fishery Management Organizations* (Washington, DC: WWF International and TRAFFIC International, 2006), 12.

manage the stock and increase the odds of attempts to free ride. Even though society in the aggregate benefits in the long run from successful management, having to decrease catches in a given year creates hardship for those individual fishers whose income will decrease (while their fixed costs may remain the same). Their time horizons may therefore be short: a fisher who cannot continue to pay off the bank loan on the fishing vessel cannot remain in the industry and so cares primarily about ensuring that catches this year are sufficient to pay the mortgage. Those short time horizons might lead to pressure for unsustainable levels of catches.

Uncertainty plays a major role at this stage as well. If there is some disagreement about the health of a fish stock (disagreement that may be played up by those with an interest in continuing catches), it is difficult to persuade states to agree to strict regulations that may turn out not to have been necessary. That difficulty in particular surfaces because there is little uncertainty about the suffering of those in the fishing industry if they are forced to curtail their fishing. But even if the regulators are correct that doing so is the only way to protect the fishery, the counterfactual is never possible to prove. If strict regulations are passed and the fishery is sustained, it is impossible to demonstrate that it was the regulations that made it possible or that without them the stocks would have crashed.

Once regulations are passed, they may not be fully implemented. Objections procedures, noncompliance, and nonparticipation can be major impediments to the implementation of even the weak rules that are agreed to within RFMOs. Objections procedures provide one means of avoiding implementation. In practice, these objections procedures are not used that frequently in some RFMOs. In ICCAT, there were objections to only three measures (by six states total) lodged across the nearly 250 regulations passed from 1969 through 2004,[46] although there has been a dramatic increase in objections more recently (between 2005 and 2010, there were objections to 18 different measures lodged by a total of eight states).[47]

---

46. Schiffman, *Marine Conservation Agreements*, 252.

47. ICCAT, *2006 Report for the Biennial Period 2004–5, Part II*, vol. 1 (Madrid: ICCAT, 2006); ICCAT, *2007 Report for the Biennial Period 2006–7, Part I*, vol. 1 (Madrid: ICCAT, 2007); ICCAT, *2008 Report for the Biennial Period 2006–7, Part II* , vol. 1 (Madrid: ICCAT, 2008); ICCAT, *2009 Report for the Biennial Period 2008–9, Part I*, vol. 1 (Madrid: ICCAT, 2009); ICCAT, *2010 Report for the Biennial Period 2008–9, Part II*, vol. 1 (Madrid: ICCAT, 2010); ICCAT, *Report for the Biennial Period 2010*, part I, vol. 1 (Madrid: ICCAT, 2011).

Likewise, in the 43 resolutions the IOTC passed between 1996 and 2010, only two objections were lodged.[48]

Other agreements, however, have seen a greater and more detrimental use of the objections procedure. States in NAFO objected to 85 different conservation measures (and objections were lodged by twelve different states for a total of 162 objections).[49] Most objections came from the European Union (EU), and the EU's spate of objections came generally on behalf of the Spanish and Portuguese fishing fleet and resulted in collective catches much higher than the overall TAC set by the commission.[50] In NEAFC since 1995 (when serious conservation measures were first adopted), member states have lodged 32 objections relating to 22 conservation measures, with Russia and Iceland responsible for the greatest numbers. Some of these objections have had little effect on actual catches, instead referring to how states are classified with respect to types of catch limits.[51] In other instances, they have had a significant effect, and catches in the area are much higher than are likely to be sustainable.

Moreover, even when, as in ICCAT, objections procedures are rarely used, their existence may nevertheless drive catch limits higher than recommended. Even those states willing to sacrifice fish catches in the short term for a fishery's long-term health know that it is better—both individually and collectively—for a catch limit to be adopted and fully implemented. Those states that might refuse to participate in a stringent catch limit therefore influence the catch levels upward, and states in favor of stronger regulation acquiesce because they would rather have a laxer standard that everyone implements than a stronger one that only some states do.

---

48. Schiffman, *Marine Conservation Agreements*, 77; IOTC, *Report of the Thirteenth Session of the Scientific Committee* (Victoria, Australia: IOTC, 2011).

49. Schiffman, *Marine Conservation Agreements*, 55; NAFO, *Meeting Proceedings of the General Council and Fisheries Commission for 2006/2007* (Dartmouth, Canada: NAFO, August 2007); NAFO, *Meeting Proceedings of the General Council and Fisheries Commission for 2009/2010* (Dartmouth, Canada: NAFO, 2010).

50. Schiffman, *Marine Conservation Agreements*, 57.

51. Ibid., 73–74; NEAFC, *Report of the 27th Annual Meeting of the North-East Atlantic Fisheries Commission, 10–14 November 2008*, vol. I (London: NEAFC, 2008); NEAFC, *Report of the 28th Annual Meeting of the North-East Atlantic Fisheries Commission, 8–12 November 2010* (London: NEAFC, 2010).

Noncompliance

Noncompliance with both domestic and international fishery rules is legion.[52] There are many different types of noncompliance: they are frequently conflated as illegal, unreported, and unregulated (IUU) fishing, although it is useful to disaggregate these categories because they represent fishing that takes place outside of the regulatory system in different ways. Illegal fishing is conceptually clearest: it involves fishing in ways that explicitly contravene rules to which the fisher is unquestionably legally bound. Unreported fishing is that which takes place behind the scenes: fishers may or may not be held to international rules, but they catch fish that they do not report to their governments or to a relevant RFMO. Their catch might be illegal, but it might not be. Unregulated fishing is the most complicated of the three concepts; fishers use a variety of legal loopholes to avoid regulation. Their catches are technically not illegal, but they circumvent the spirit of the existing rules.

Studies have shown that fishers who doubt the effectiveness of fishery regulations in protecting the resource are less likely to comply with these rules,[53] thereby creating a catch-22 because their very noncompliance contributes to undermining these rules' effectiveness.

Flag-of-convenience (FOC) registration, in which vessel owners register their ships in and thereby acquire the nationality of states that do not impose strict regulations,[54] has been a useful way for fishing vessels to avoid participating in RFMOs. As a result, there is much fishing outside of the regulatory process in many of the most important regulatory regions. At least 10 percent of the world's fishing vessels are registered under FOCs, with newly built vessels even more likely than existing ones

52. Christopher C. Joyner, "Compliance and Enforcement in New International Fisheries Law" *Temple International and Comparative Law Journal* 12 (Fall 1988): 271–299; Christopher J. Carr, "Recent Developments in Compliance and Enforcement for International Fisheries," *Ecology Law Quarterly* 24 (1997): 847–860.

53. Jesper Raakjaer Nielsen and Christoph Mathiesen, *Incentives for Compliance Behaviour: Lessons from Danish Fisheries*, Institute for Fisheries Management and Coastal Community Development Report no. 68 (Exeter, UK: Institute for Fisheries Management and Coastal Community Development, 2001), http://www.ifm.dk/reports/68.pdf.

54. Elizabeth R. DeSombre, *Flagging Standards: Globalization and Environmental, Safety, and Labor Standards at Sea* (Cambridge, MA: MIT Press, 2006).

to register in these locations.[55] Some estimates suggest the real number might be as high as 21.5 percent,[56] and FOC-registered vessels are likely to be larger than average.[57] Most RFMOs experience high levels of fishing by vessels registered in nonmember states. Both ICCAT and the IOTC have estimated that about 10 percent of the catch in their regulatory areas is taken by fishers with vessels registered in nonmember states.[58] The CCSBT estimates its nonmember state catch to be between 15 and 35 percent.[59] A variety of estimates put the percentage of fish caught by nonmember-flagged vessels in CCALMR's regulatory area at between two and four times the amount of regulated fishing.[60]

Another way that regulations may not have their intended outcome involves interaction effects. Most commercial fishing operations focus on a particular species of fish without regard for the broader ecosystem in which that species functions. Most RFMOs regulate this way as well. But fish eat other fish and are eaten in turn by bigger fish. Focusing on just one species ignores the way other species in the ecoystem are affected by the depletion of the target species and the way in which the target species might in turn be affected by what is happening with other species.

RFMOs have more recently begun to consider the broader ecosystemic interaction effects of regulations. CCAMLR is the first one to take up this

55. DeSombre, "Fishing under Flags of Convenience."

56. International Confederation of Free Trade Unions, Trade Union Advisory Committee to the OECD, ITC, and Greenpeace International, "More Troubled Waters: Fishing Pollutions and FOCs," Major Group Submission for the 2002 World Summit on Sustainable Development, Johannesburg.

57. Matthew Gianni and Walt Simpson, *Flags of Convenience, Transshipment, Re-supply, and at-Sea Infrastructure in Relation to IUU Fishing*, OECD, Fisheries Committee, Directorate for Food, Agriculture, and Fisheries, AGR/FI/IUU (Rome: FAO, 2004).

58. OECD, *Draft Chapter 2—Framework for Measures against IUU Fisheries Activities*, AGR/FI/IUU(2004)5/PROV (Paris: OECD, Fisheries Committee, Directorate for Food, Agriculture, and Fisheries, 2004).

59. Judith Swan, *Fishing Vessels Operating under Open Registers and the Exercise of Flag State Responsibilities—Information and Options*, FAO Fisheries Circular no. 980, FITT/C980 (Rome: FAO, 2002).

60. CCAMLR, *Explanatory Memorandum on the Introduction Catch Documentation Scheme (CDS) for Toothfish*, n.d., http://www.ccamlr.org/pu/E/cds/p2.htm; M. Lack and G. Sant, "Patagonian Toothfish: Are Conservation and Trade Measures Working?" *TRAFFIC Bulletin* 19 (1) (2001): 1–18.

issue explicitly.[61] The idea is that even when single-species catch limits are set, the commission considers not only that species' sustainable yield, but its effect on others, ensuring that a sufficient number remain, for instance, to serve as prey for other species.[62] Although most agree that CCAMLR's new approach is important conceptually,[63] the extent to which the commission's regulations are actually affected by consideration of broader ecosystem effects is less clear.[64]

Another problem is that of bycatch—fish that are unintentionally caught when a particular species is being fished. Commercial fishing operations discard between 18 and 40 million tons of nontarget species annually.[65] The FAO has estimated that up to 25 percent of global fish catch is bycatch.[66] A more recent estimate using an expansive definition estimated bycatch at 40.4 percent of marine fish catches.[67] Even if bycatch is returned to the ocean, the mortality is high (up to 100 percent), depending on the fishing method. Some methods, such as trawling, that produce the greatest levels of bycatch also produce the greatest mortality in bycatch.[68] From a scientific perspective, an additional problem of bycatch is that neither the discards nor their level of mortality is generally

---

61. Convention for the Conservation of Antarctic Marine Living Resources (1980), Article IX(2)(i).

62. Andrew J. Constable, William K. de la Mare, David J. Agnew, Inigo Everson, and Denzil Miller, "Managing Fisheries to Conserve the Antarctic Marine Ecosystem: Practical Implementation of the Convention on the Conservation of Antarctic Marine Living Resources (CCAMLR)," *ICES Journal of Marine Science* 57 (2000): 778–791.

63. See, for example, Mary Ruckleshaus, Terrie Klinger, Nancy Knowlton, and Douglas P. DeMaster, "Marine Ecosystem-Based Management in Practice: Scientific and Governance Challenges," *BioScience* 58 (1) (2008): 53–63.

64. Karl-Hermann Kock, Keith Reid, John Croxall, and Stephen Nicol, "Fisheries in the Southern Ocean: An Ecosystem Approach," *Philosophical Transactions of the Royal Society, Biological Sciences* 362 (1488) (2007): 2333–2349.

65. David L. Alverson, Mark H. Freeberg, Steven A. Murawsk, and J. G. Pope, *Global Assessment of Fisheries Bycatch and Discards*, FAO Fisheries Technical Paper no. 339 (Rome: FAO, 1994).

66. Ivor Clucas, *Fisheries Bycatch and Discards*, FAO Fisheries Circular no. 928 (Rome: FAO, 1997).

67. R. W. D. Davies, S. J. Cripps, A. Nicksona, and G. Porter, "Defining and Estimating Global Marine Fisheries Bycatch," *Marine Policy* 33 (2009): 661–672.

68. S. Pascoe, *Bycatch Management and the Economics of Discarding*, FAO Technical Paper no. 370 (Rome: FAO, 1997).

recorded.[69] The lack of a record for bycatch means that scientific efforts that draw conclusions based on effects of recorded commercial catches are dramatically undercounting the actual fishing effort taking place. The extent of depletion may be much greater than expected for a given catch limit, and recovery for depleted fisheries will take longer than predicted.

A related issue is "high grading"; fishers with quotas that limit the amount of fish they can land want to bring in the fish that will have the highest value on the market. They may therefore discard already-caught fish of the regulated species if they catch higher-quality fish later. The health of the discarded fish is questionable at best. Most fish discarded in this manner do not survive.[70] High grading can happen even when no catch limits are imposed externally, but physical limitations such as space or cooling or freezing capacity exist.[71]

The RFMO Concept

The bigger conceptual problems with RFMOs is that the way they divide the ocean and ecosystem across categories makes sense neither scientifically nor politically, even though these categories may have originated for both scientific and political reasons. The earliest RFMOs formed in areas where overfishing looked as if it was becoming a problem, but also—most important—where the primary states involved recognized the need. The United States and Canada established the International Fisheries Commission in 1923 (it later became the International Pacific Halibut Commission).[72] Once the first few RFMOs had formed, others followed, often using similar approaches.

---

69. R. Hilborn and C. J. Walters, *Quantitative Fish Stock Assessment: Choice, Dynamics, and Uncertainty* (New York: Chapman and Hall, 1992).

70. Frank Chopin, Yoshihiro Inoue, Y. Matsushita, and Takafumi Arimoto, "Source of Accounted and Unaccounted Fishing Mortality," in *Solving Bycatch: Considerations for Today and Tomorrow*, ed. Alaska Sea Grant College Program, 41–47, Report no. 96-03 (Fairbanks: University of Alaska, 1996).

71. Ragnar Arnason, "On Catch Discarding in Fisheries," *Marine Resource Economics* 9 (3) (1994): 189–207; Ragnar Arnason, *On Selectivity and Discarding in an ITQ Fishery: Discarding of Catch at Sea*, Working Paper no. 1 (Reykjavik: Department of Economics, University of Iceland, 1996).

72. International Pacific Halibut Commission, "About IPHC," n.d., http://www .iphc.washington.edu/about-iphc.html.

Even within RFMOs, some of the additional designations are politically derived. A case in point is the separation of management within ICCAT
into eastern and western stocks, each of which is subject to its own quota
system. The process of managing populations separately began in 1981.
Tagging and genetic studies have since demonstrated, however, that the
mixing between the two "stocks" is high, with some arguing that the species should be managed as one single stock.[73] But managing the stock in
a unified way would challenge vested interests that have figured out how
best to work within the current divided approach, so there is no political
support for what is ecologically the best management practice.

There has recently been a trend toward cooperation across RFMOs
whose mandates overlap. The five different regional organizations managing tuna and tunalike species have begun periodic collective meetings and
have agreed to harmonize data collection and to improve coordination on
issues such as scientific research, market analysis, bycatch prevention, and
efforts to prevent fishing outside of these organizations' regulatory processes.[74] The tuna RFMOs, as a part of this new cooperative process, have
also begun a process of self-study, prompted by the difficulties they have
experienced in successfully managing the fisheries they oversee.[75] Other
RFMOs, such as NEAFC, have also undertaken some kind of externally
conducted performance review.[76]

The current RFMOs nevertheless still make management decisions
separately by species and by region. The biggest problem that results
from this piecemeal approach to global fisheries regulation is the balloon
problem, discussed in greater detail in chapter 3. As long as global fishing

73. Commission on Geosciences, Environment, and Resources, *An Assessment of
Atlantic Bluefin Tuna* (Washington, DC: National Academies Press, 1994); Clay
E. Porch, "The Sustainability of Western Atlantic Bluefin Tuna: A Warm-Blooded
Fish in a Hot-Blooded Fishery," *Bulletin of Marine Science* 76 (2) (2005): 363–384.

74. "Draft Report of the Second Joint Meeting of Tuna Regional Fisheries Management Organizations (RFMOs)," 2nd Joint Tuna RFMOs Meeting, San Sebastian, Spain, 29 June–3 July 2009.

75. "Report of the Tuna RFMOs Chairs' Meeting," San Francisco, 5–6 February 2008, http://www.tuna-org.org/Documents/RFMO_CHAIRS_FEB%205-6_
FRISCO_Phils.pdf.

76. NEAFC, Performance Review Panel, *Report of the North East Atlantic Fisheries Commission: Performance Review Panel Report of the NEAFC* (London:
NEAFC, 6 November 2006), http://www.neafc.org/system/files/performance-
review-final-edited.pdf.

effort does not decrease, regulations for one species or in one region simply move the fishing effort elsewhere. It is this problem that most urgently needs to be addressed in order to protect global fisheries and the ecosystems of which they are a part.

The other underlying problem is conceptual. Despite their political origins and organization, RFMOs at base operate with a scientific logic: the idea that the sustainable yield of a fish stock can be determined and that it is this scientific estimate that is the most relevant for regulatory purposes. When criticism is made of the use of science within these organizations, it most frequently argues that the science is too uncertain or that RFMOs do not follow the scientific advice they are given rather than that the whole edifice of scientific advice is problematic. The problem, however, is that RFMOs are designed, through scientific advice, to conserve fish; they therefore focus on how to limit catches of the specific species they regulate in the region for which they are responsible. But by omitting a broader focus on the sociopolitical context of the fishers and the fishing industry, they are not able to get to the underlying problems that threaten regional and global fish stocks.

Conclusions

Ocean fisheries are difficult to manage under the best of circumstances. Fish exist in or pass through international spaces, where no overarching authority has the ability to impose regulation for their protection. It is both legally and practically difficult to prevent people from gaining access to fisheries, and the more fish that are taken, the fewer that remain to be taken. Those incentives set up a potential "tragedy of the commons" in which there are benefits to taking advantage of fishery resources in the short run rather than conserving them for the long run, especially if there is no certainty that the resource will be well managed over time.

The organizations undertaking regionally based management efforts have done their best to create agreements in some places to protect some species. RFMOs attempt to study fish stocks and to create and enforce rules designed to fish them at a level that will allow fishing to continue indefinitely. And although uncertainty and issues of participation and compliance make this regulatory process difficult, the bigger underlying problem is that regulating by region and species is problematic in an industry where many of the actors operate globally.

3

The Political Economy of Regulation

RFMOs are designed explicitly to deal with the CPR aspect of international fisheries. They are supposed to take an open-access resource and make it partially excludable by making states limit the amount of the resource their fishers have a right to remove from the commons. And yet in the aggregate RFMOs are not generating effective global fisheries governance.[1] The concept of a CPR is one that draws heavily on the concepts of the economics discipline, and economics provides analytical tools for addressing the management problems that CPRs present. The field of environmental economics in particular was developed to address the market imperfections of resource use and pollution.[2] RFMOs recognize that international fisheries are a CPR and often explicitly draw on the analytic tools of environmental economics to generate more effective fisheries governance. At the same time, they for the most part apply a scientific rather than an economic logic, one drawn largely from population biology, to determine how much can be fished and how it can be done. Why are the analytical tools of environmental economics and population biology in combination not more effective in generating better management?

Part of the answer is that politics interferes with the application of these tools. RFMO secretariats propose rules based on scientific analysis of fish stocks and economic analysis of the potential for effective

---

1. See, for example, Sarika Cullis-Suzuki and Daniel Pauly, "Failing the High Sea: A Global Evaluation of Regional Fisheries Management Organizations," *Marine Policy* 34 (5) (2010): 1036–1042.

2. Thomas H. Tietenberg, *Environmental and Natural Resource Economics*, 5th ed. (Reading, MA: Addison-Wesley, 2000).

implementation, but country representatives, who have final decision-making powers, overrule their advice. Lack of political will on the part of countries and their representatives to deal with the international fisheries crisis is indeed a real problem, to which chapter 4 is devoted. In a way, however, this answer only begs another question. Country representatives who participate in undermining secretariat proposals are acting as classic free riders, and overcoming the free-rider problem is one of the things that environmental economics should be able to help address.[3] And another part of the problem is the various kinds of uncertainty that plague fisheries regulation. But a further and distinct problem lies in the tensions between the logic of CPR analysis and the logic of population biology as they have been applied to international fisheries management.

The latter problem has three parts. The first has to do with the different ways environmental economists and population biologists approach resource management in general and international fisheries management in particular. At its simplest, environmental economics is about interacting with the natural environment in a way that maximizes present economic value. Multilateral environmental cooperation is not, however, focused purely on an econocentric worldview. It explicitly invokes an environmental sustainability perspective as well. The scientific models that underlie fisheries regulations are generally premised on maintaining steady yield over time rather than present value. The tension between these perspectives has negative ramifications for fisheries management.

The second part of the problem with the way CPR analysis has been applied to fisheries issues is that the basic CPR model is an insufficient starting point for analysis. Markets do not succeed on their own to regulate the exploitation of CPRs because of what economists call a market failure. There are multiple kinds of market failure, and the logic of CPRs focuses on one of them, the absence of clear property rights.[4] Fisheries management suffers from this problem, and most RFMO rules are designed to address it. But fisheries present another kind of market failure as well, the absence of information—in particular the absence of information about when fish stocks will fail.

---

3. John McMillan, "The Free-Rider Problem: A Survey," *Economic Record* 55 (2) (1979): 95–107.

4. Tietenberg, *Environmental and Natural Resource Economics*, 68–71.

Standard CPR analysis, furthermore, is predicated on the assumption that there is one tier of economic actors and of governance.[5] Individuals or corporations use resources and either cooperate to manage them or have management imposed by government (or some combination thereof). But international fisheries management is more complicated. States (or fisheries bureaucracies within states) are both managers of the resource at the national level and the equivalent of users of the resource at the international level. RFMOs generally operate under the assumption that states (or at least member states) will act as rational users from the perspective of international management. But the multi-level dynamics of the international politics of CPR management are not well understood.

The third part of the problem with the way CPR analysis has been applied to fisheries issues has to do with the actual original users of international fishery resources, the fishers. The classic treatment of CPRs is as a closed system—there is one resource, and there are users who need to be prevented from overexploiting it.[6] This treatment assumes that the users either will make do with regulated levels of resource exploitation or will exit the system and find another livelihood. But there are barriers to exit from the fishing industry, so that, where possible, fishers may be more likely to try to relocate within the industry than to leave it. Furthermore, fisheries—in particular international fisheries—are not as closed a system as in the basic CPR model. The industry involves many kinds of fishers who are exploiting many different fish stocks. Regulating any aspect of this system as a CPR fails to address the interactions among parts of the industry. This approach to regulation creates what we are calling a balloon problem—effective regulation of one part of the industry may have the effect of increasing pressure on other parts. In this sense, RFMO management will not scale up because, even when effective, it displaces rather than solves the problem of overfishing, unless barriers to exit from the industry are removed.

---

5. See, for example, Elinor Ostrom, Roy Gardner, and James Walker, *Rules, Games, & Common-Pool Resources* (Ann Arbor: University of Michigan Press, 1994).

6. For example, H. Scott Gordon, "The Economic Theory of a Common-Property Resource: The Fishery," *Journal of Political Economy* 62 (2) (1954): 124–142. Although many scholars have modified the Gordon model, little attention has been paid to the possibility of a balloon problem.

Environmental Economics

The standard first response from within economics to the problem of common access that is the source of the overexploitation of CPRs is to limit access. Doing so requires agreement among the users of the resource and some form of group enforcement, or government fiat and enforcement, or the privatization of the commons. The middle option, government directive, is not viable in international politics because there is no central government capable of limiting access to a global resource. Nor does international law allow for regulation to apply to countries that have not explicitly consented to it.[7] Efforts at governance of global fisheries have therefore focused on the other two options. The single most ambitious change in global governance of fisheries was based on the third of the options, privatization. The creation of 200-mile EEZs, begun in the middle of the twentieth century and formalized in UNCLOS III, put some 90 percent of existing living marine resources within national jurisdiction.[8] This scheme constitutes privatization in an international legal sense, in which states are treated as individual actors. But in many cases this change did not yield the expected and desired effect of better management of fisheries resources, for reasons that are addressed in chapter 4.

RFMOs represent the first of the options, agreement among users to limit use of the resource. Governance through this mechanism faces the variety of obstacles discussed in chapter 2, including such things as monitoring and enforcement, the same obstacles faced by international cooperation in many issue areas. This model of cooperation assumes a general agreement on goals. In the case of the classic example of the tragedy of the commons, livestock grazing on a village commons, the goal of cooperation is maximizing the villagers' well-being.[9] It so happens that their well-being is generally tied to the livestock's well-being. This confluence is by coincidence, however, not by design. Environmental economics sees CPR management as a problem of human economic maximization but

---

7. Daniel Bodansky, *The Art and Craft of International Environmental Law* (Cambridge, MA: Harvard University Press, 2010).

8. David A. Colson, "Current Issues in Fishery Conservation and Management," *U.S. Department of State Dispatch* 6 (7) (1995), http://dosfan.lib.uic.edu/ERC/briefing/dispatch/1995/html/Dispatchv6no07.html.

9. Garrett Hardin, "The Tragedy of the Commons," *Science* 162 (1968): 1243–1248.

does not see the maintenance of the resource for its own sake as a legitimate goal in its own right.[10] But RFMOs are also bound by international norms of stewardship, precaution, and sustainability.[11]

The anthropocentric and econocentric norms of environmental economics can come into tension with those of stewardship, precaution, and sustainability.[12] To put it simply, these latter perspectives put a premium on the value of maintaining ecological systems and biodiversity in the long term. The econocentric perspective, in contrast, attaches no value to biodiversity and robust ecological systems per se unless they yield human utility that can be expressed as a monetary value. In the example of the village commons, this distinction has no impact because the assumption is that maintaining the existing ecosystem (which is a pastoral rather than a natural system) will maximize economic utility. But the coincidence of economic and ecological utility is not necessarily going to hold true in fisheries management.

Two examples illustrate this point. One example contrasts the economic value of an ecologically mature ecosystem with that of a much simpler ecosystem. The Grand Banks cod fishery was historically one of the world's richest, but it was overexploited in the 1970s and 1980s to the point of complete collapse. The Canadian government declared a moratorium on the fishery to allow it to recover.[13] But the collapse of the cod stock was so complete that cod's ecological niche was taken over by shrimp in a less biodiverse and more primitive ecosystem. It turned out, however, that the shrimp are a more lucrative fishery than the cod they replaced. The collapse of the cod fishery from overexploitation had definite negative ecological and social consequences, but in purely economic terms it may have generated a net utility gain.[14]

---

10. J. Samuel Barkin, "Discounting the Discount Rate: Ecocentrism and Environmental Economics," *Global Environmental Politics* 6 (4) (2006): 56–72.

11. In some cases, such as CCAMLR, these norms can be found in the organization's founding documents. For other RFMOs, such as ICCAT, the founding documents focuses entirely on MSY.

12. On the relationship between environmental economics and sustainability, see Barkin, "Discounting the Discount Rate."

13. Mark Kurlansky, *Cod: A Biography of the Fish That Changed the World* (New York: Walker, 1997).

14. Alex Rose, *Who Killed the Grand Banks? The Untold Story behind the Decimation of One of the World's Greatest Natural Resources* (Mississauga, Canada: Wiley, 2008).

A second example can be found with particularly slow-growing species such as orange roughy. Environmental economics maximizes the current value of resource use by discounting future values. In other words, a catch of a given size next year has a lower present value than a catch of the same size this year, and a catch in, say, ten years has an even lower present value.[15] The cumulative effect of future catches often means that, despite this discounting, it makes sense to forego a big catch now in order to maintain a viable fishery in the future, but that forbearance is not always economically rational. If a species takes decades to reproduce, it may make more sense from an economic perspective to catch all that can be caught now and forego the future benefits of a viable fishery several decades hence.[16] Such a decision might make sense from an environmental economics perspective but not from a sustainability or stewardship perspective. The tensions among these perspectives have the effect of muddling RFMOs' goals.

Discounting future values can undermine global fisheries governance as a practice and as an analytical tool. To the extent that states are thinking about their roles in RFMOs in economic or development terms, they are likely in practice to discount the future, valuing economic benefits in the near term more than those in the distant future. The extent to which they do so, however, differs across countries. These differences can be systematic (other things being equal, poor countries are likely to discount the future more than rich ones) or contextual. The greater the extent to which a country discounts the future with respect to a particular resource, the greater its bargaining power in international negotiations with respect to that resource, other things being equal. Countries that heavily discount the future can credibly threaten to use up a CPR if negotiations do not succeed, thus putting pressure on countries that value the future more highly to come to an agreement on terms that favor those that value the future less.[17]

In other words, those countries that care the least about the long-term health of a resource are best placed to set the terms of international negotiations about its governance. This advantage can be (and in many cases

---

15. Anthony Scott, "The Fishery: The Objectives of Sole Ownership," *Journal of Political Economy* 63 (2) (1955): 116–124.

16. Colin W. Clark, "The Economics of Overexploitation," *Science* 181 (4100): 630–634.

17. J. Samuel Barkin, "Time Horizons and Multilateral Enforcement in International Cooperation," *International Studies Quarterly* 48 (2) (2004): 363–382.

is) overcome by countries that discount the future less and threaten either political escalation of fisheries disputes or the use of market power to punish countries that threaten to undermine resource governance.[18] But the basic balance of power in fisheries negotiations when these negotiations are driven by an environmental economics logic favors short-term use rather than long-term stewardship of fisheries resources. Only by embedding negotiations in the social logics of precaution and sustainability can this basic balance be altered.

Most RFMOs are in fact committed to principles of sustainability, implying that they are committed to maintaining the health of fish stocks over time. One of the concepts used most often in existing fisheries management is that of MSY, even though it is not accepted as unproblematic by either environmental economists or population biologists.[19] This concept seems at first to fit with norms of stewardship—after all, it has the word *sustainable* in it. The concept of MSY comes from population biology and is based on the idea that current catch levels should be set in a way that maximizes catch levels over time. Although this concept does not allow for discounting the future, as an environmental economics approach does, it is still a fundamentally anthropocentric perspective. Its core measure remains the resource's human value (in terms of edible protein over time rather than present economic value), not its ecological health.[20]

The concept of MSY ironically often calls for higher levels of fishing than the equivalent economic measure, maximum economic yield (MEY), because the economic measure takes the cost of fishing into account,

---

18. J. Samuel Barkin and Elizabeth R. DeSombre, "Unilateralism and Multilateralism in International Fisheries Management," *Global Governance* 6 (3) (2000): 339–360.

19. For example, see Donald Ludwig, Ray Hilborn, and Carl Walters, "Uncertainty, Resource Exploitation, and Conservation: Lessons from History," *Science* 260 (1993): 17, 36.

20. The concept can be seen as a sort of compromise between economic and ecological logic that fully conforms with neither. It seems to be used as often as it is at this point as much because of institutional momentum as anything else. See, for example, Willard Barber, "Maximum Sustainable Yield Lives On," *North American Journal of Fisheries Management* 8 (2) (1988): 153–157, and Christopher Corkett, "The PEW Report on US Fishery Councils: A Critique from the Open Society," *Marine Policy* 29 (3) (2005): 247–253. For an interesting complication of maximum yield concepts, see Ray Hilborn, "Defining Success in Fisheries and Conflicts in Objectives," *Marine Policy* 31 (2) (2007): 153–158.

whereas MSY does not.[21] And the fact that MSY often calls for higher levels of fishing than does economic calculation can help to drive up levels of subsidization of the fishing industry, as discussed in chapter 6. When fishers are authorized to catch more than they can do profitably, they can do so only if subsidized.

There are several other ways MSY can undermine ecological as well as economic value and even undermine the goal of maximizing food inputs over the long term. One ecological effect is on the specific species being fished at MSY. Over time, even if fishing at MSY allows for the maintenance of the size of the overall catch, it may well affect the characteristics of average individuals within the species. It will have a tendency to make them smaller and shorter-lived.[22] An evolutionary logic prevails: if there is intensive fishing happening in an area, fish that best survive in a population will be those that mature the fastest because they are the ones that can reproduce before they are caught. Those fish will then be favored in the next generation and are likely to be shorter-lived because of their earlier reproductive maturity. At the same time, smaller fish at a given age are more likely to escape being caught in a fishing operation and will thus be the ones to survive to produce the next generation. These species-specific effects in turn may impact the ecosystem in which the species lives.

In addition, by significantly decreasing the total living biomass of a species, MSY can have significant impacts on ecosystems. MSY-caused reductions in biomass mean that the species as a whole will eat less, making room for other predators. This change can affect the system's overall ecological balance in unpredictable ways. Reduction in biomass also means less food for whatever other species eats the target species, whether the former are species caught in other commercial fisheries or are species such as seabirds that are noncommercial.[23]

---

21. For example, see Colin W. Clark, *Mathematical Bioeconomics: The Optimal Management of Renewable Resources*, 2nd ed. (New York: Wiley-Interscience, 1990). In *The Worldwide Crisis in Fisheries* (Cambridge: Cambridge University Press, 2006), Clark criticizes MEY as well, but, interestingly, he does so because he argues that it leads to political failure.

22. John Sibert, John Hampton, Pierre Kleiber, and Mark Maunder, "Biomass, Size, and Trophic Status of Top Predators in the Pacific Ocean," *Science* 314 (2006): 1773–1776.

23. See, for example, Carl Walters, Villy Christensen, Steven Martell, and James Kitchell, "Possible Ecosystem Impacts of Applying MSY Policies from Single-Species Assessment," *ICES Journal of Marine Science* 62 (3) (2005): 558–568.

Finally, MSY assumes more knowledge of fish biology and ecology than is often warranted and does not sufficiently account for natural variability and natural change in ecosystems.[24] The calculation of MSY requires accurate knowledge of lifespan and breeding patterns, which marine scientists cannot necessarily provide. The more recently a particular species became commercially important, the less likely it is that we have this information because much of the data we have on many fish stocks comes from commercial fisheries. This means that new fisheries that are exploited in response to increased regulation on existing fisheries are likely to be less well managed than those they replace. Natural variability also provides a real challenge to fisheries management of any kind. We do not have enough information on the vast majority of stocks to know how much natural variability there is over time because accurate records do not go back far enough. Allowing for this uncertainty, per the precautionary principle, would require fishing at well below MSY.

A Complicated CPR

Even from just the environmental economics perspective, without the complicating relationship with population biology, global fisheries governance constitutes a particularly tricky CPR problem. The resource itself, fish stocks, has different characteristics than other CPRs, and global governance of resources (unlike domestic governance) is a two-level game. These factors are problematic because they generate complex and uncertain inputs and therefore do not fit well with the calculations that drive most policy-related models, whether they be economic or biological. These models are designed to yield precise results, in particular stable equilibria. Complex and uncertain inputs undermine this precision and mean that equilibria, whether economic or biological, are less likely to be stable.[25] Given these complex and uncertain inputs, successful fisheries governance requires a much more precautionary approach than would be the case for a more straightforward CPR.

There are several major sources of complexity and uncertainty in inputs to biological models of fish recruitment (the number of new fish that

24. Pamela Mace, "A New Role for MSY in Single-Species and Ecosystem Approaches to Fisheries Stock Assessment and Management," *Fish and Fisheries* 2 (2001): 2–32.

25. Clark, *The Worldwide Crisis in Fisheries.*

survive to join the fish population) and therefore to maximum yield models of fisheries. One is the relationship between fish populations and fish reproduction. A simple model would suggest a decline in recruitment equivalent to a decline in population beyond the point at which yields are maximized. Such a model would give some warning of overexploitation—stocks would show clear signs of decline as they were overfished. But some species do not appear to behave in this way—they show a constant return to fishing effort up until a particular point of overexploitation is reached, after which there is a stock collapse that is rapid enough that even major cuts in quotas cannot stem it.[26] Scientists are getting better at modeling these relationships, but it remains difficult to do so accurately. This problem is exacerbated by the fact that we often do not have accurate information about stock sizes. The information that we do have is often derived from information about how hard fish are to catch (i.e., fishing effort required), which is not necessarily directly related to how abundant they are.[27]

Another major source of complexity is the relationships among species in an ecosystem. If, for example, the population of a species is reduced through fishing, it may allow more room for recruitment within the species, as competition for food decreases. But it may also allow other species to take advantage of a niche with less competition. Furthermore, the effects of such a reduction on other species for which the reduced species is prey are often unknown. This complexity is similar to the ecosystemic problems with MSY discussed earlier: ecosystems are more complex than we can accurately model with our existing knowledge. The third major source of complexity and uncertainty is natural variation due to cyclical weather patterns or secular weather changes, long-term population dynamics within the species, or other natural sources of change. This source of uncertainty will likely be exacerbated by increasing rates of climate change in the future. Both economic and ecosystemic modelers try to take these factors into account, but doing so increases the imprecision inherent in the models.[28]

---

26. Christian Mullon, Pierre Fréon, and Philippe Cury, "The Dynamics of Collapse in World Fisheries," *Fish and Fisheries* 6 (2) (2005): 111–120.

27. See Clark, *The Worldwide Crisis in Fisheries*, and Carl J. Walters and Steven J. D. Martell, *Fisheries Ecology and Management* (Princeton, NJ: Princeton University Press, 2004), for more detailed discussion of both these problems and ways in which ecological and economic modelers are dealing with them.

28. Clark, *The Worldwide Crisis in Fisheries*; Walters and Martell, *Fisheries Ecology and Management*.

As well as uncertain science, global fisheries governance suffers from uncertain politics. International fisheries represent a two-level CPR. The standard model of CPRs has one level of politics. If a management model is agreed upon collectively, or if one is imposed authoritatively, then the users of the resource can be expected to act on it as individuals, deciding either to comply or not to comply. Enforcement may then present a free-rider problem, but that is an ancillary issue. With international fisheries, the parties negotiating the management models are states, not fishers. When states come to agreement to manage international fishery resources, either through privatization or cooperation on regulations and quotas, they then acquire a domestic CPR problem. Whether by managing access to coastal waters the states have come to control or managing access to a quota that they are responsible for distributing, successful management of the international CPR problem in turn requires that all the various relevant domestic CPR problems be managed successfully as well.

Robert Putnam identified this two-level model of international politics in a negotiations context,[29] which helps to explain some of the difficulties (examined further in chapter 4) when state representatives in RFMOs have to navigate their domestic actors' complex preferences and may intentionally bargain in a way that plays the domestic and international levels against each other. But this two-level model also pertains to the actual management context as well, helping to explain how states behave domestically when implementing international rules.

The national-level CPRs that result from international cooperation should in principle be well managed. After all, this cooperation gives states valuable resources in the form of either a newly domestic fishery or a valuable quota. States should then recognize that sound management will maintain the value of that resource into the future. But things often do not work out that way. There is the issue of enforcement. As is the case with a single-level CPR, enforcement can create a free-rider problem. In the single-level case, the free-rider problem concerns the policing of others' behavior. This problem is significant at both the domestic level and the international level and is made more complicated at the international level by the absence of a central authority able to enforce international agreements.

---

29. Robert Putnam, "Diplomacy and Domestic Politics: The Logic of Two-Level Games," *International Organization* 42 (3) (Summer 1988): 427–460.

The free-rider problem is more complicated at the international level because the resource in question is often not as important to the international-level actors, states, as it is to the key actors in domestic CPR issues. For example, maintaining a healthy commons for grazing was a matter of life or death for medieval peasants, and one can therefore expect that they were willing to prioritize enforcement of commons agreements over most other issues. Maine lobster fishers are similarly willing to go beyond the law to enforce access rights to their resource because this access is necessary to their lifestyle and livelihood.[30] But with very few exceptions, fishing is not economically vital to national economies, even—or especially—to the big wealthy economies that subsidize their fleets most heavily. For example, the value of fish landed in the United States and the United Kingdom is equivalent to one-thirtieth of one percent of gross domestic product (GDP). For countries such as Canada and Spain, the equivalent figure is one-sixth of one percent, and for Japan it is one-quarter of one percent. The exceptions are countries with small populations and large coastlines.[31] The value of fish landed in Norway, for example, is equal to three-quarters of one percent of GDP, and the equivalent figure for Iceland is more than 9 percent.[32] Except in these latter locations, therefore, states should be less likely to be willing to expend political capital with other states in order to enforce fishing agreements than they are for issues such as national security. They are also less likely to be willing to sacrifice for these relationships than are peasants or lobster fishers.

Beyond the added complications of the enforcement free-rider problem at the international level, there is also the problem of self-policing. With a single-level CPR, self-policing is not an issue—individual users of a resource can be expected to know what they are doing and to enforce their own decisions on themselves (although the situation may be more complicated with corporate users). In other words, if I have committed to catching only a certain amount of fish, I generally know whether I have caught that amount or not.

---

30. See, for example, James M. Acheson, *Capturing the Commons: Devising Institutions to Manage the Maine Lobster Industry* (Lebanon, NH: University Press of New England, 2003).

31. These countries ironically subsidize their fishers much less heavily.

32. Landing figures are from OECD, *Review of Fisheries in OECD Countries 2009: Policies and Summary Statistics* (Paris: OECD, 2010), 49. GDP figures are from OECD, *OECD Factbook 2010* (Paris: OECD, 2010), 33. Figures in both cases are for calendar year 2006.

In the case of many forms of international cooperation, however, states are the users at the international level (with a state-level fishing quota) but the enforcers at the domestic level. States have to provide the (international-level) public good of policing domestically. If a state has committed to having its fishers catch only a certain amount of fish, the "state" may not actually know, without active effort, how much fish its substate actors are catching. Such policing can be costly, both financially and, perhaps more important, politically,[33] because it may require picking fights with fishing industry lobbies. Two-level CPR cooperation thus suffers not only from intentional cheating, but also from unwillingness to enforce agreements domestically even when governments would prefer, absent domestic political costs, to comply with their commitments.

This sort of two-level-game problem affects many forms of international regulatory cooperation, especially those that require countries in turn to regulate private business domestically.[34] It is particularly acute in the case of international fisheries regulation because this issue area is one of the few that deals both with a CPR and with the regulation of nationals extraterritorially. The CPR aspect means that those countries that fail to enforce rules domestically can actively undermine other countries' enforcement efforts by allowing fish stocks to be overexploited. And the extraterritorial aspect means that unregulated fishers are operating in the same place as regulated fishers, making it more difficult to isolate noncooperators. It also makes it easier for fishers to avoid regulation by states that enforce rules effectively.[35] The FOC problem discussed in chapter 2 originates from this difficulty.

Some fish stocks' tendency to hold steady over time in the face of overexploitation until they suddenly dramatically decrease and the

33. Robert Keohane, Peter Haas, and Marc Levy argue that domestic capacity constraints are one of the key hindrances to effective international environmental cooperation. See Robert O. Keohane, Peter M. Haas, and Marc A. Levy, "The Effectiveness of International Environmental Institutions," in *Institutions for the Earth: Sources of Effective International Environmental Protection*, ed. Peter M. Haas, Robert O. Keohane, and Marc A. Levy, 3–24 (Cambridge, MA: MIT Press, 1993).

34. Peter B. Evans, Harold K. Jacobson, and Robert D. Putnam, eds., *Double-Edged Diplomacy: International Bargaining and Domestic Politics* (Berkeley: University of California Press, 1993).

35. Elizabeth DeSombre, "Fishing under Flags of Convenience: Using Market Power to Increase Participation in International Regulation," *Global Environmental Politics* 5 (4) (2005): 73–94.

complications of a two-level politics of international regulatory cooperation make finding the right balance in international fisheries management difficult. The fish stocks' response pattern means that regulation needs to be particularly precautionary—regulators need to allow the stocks some slack rather than to aim for MSY because if they aim for the maximum, they are likely to overshoot half the time. And because stocks often give little warning before they collapse, overshooting is difficult to remedy. One cannot wait for clear signs of decline and then adjust because by then it may be to be too late to avoid collapse. Meanwhile, the political response pattern is to underenforce agreements, even when states sign those agreements without any intention of cheating. Agreements therefore need to be more stringent than demanded by science because they are not likely to be fully honored. Both of these factors suggest that fisheries regulation should be made significantly tighter than the numbers suggested by measures of maximum yield, whether MSY or MEY.

Balloons and the Surplus of Fishers

But even if individual agreements are made more stringent than demanded by MSY calculations, and RFMOs can be persuaded to adopt a more precautionary stance than they currently do, the cumulative effect of RFMO governance is still likely to be less than the sum of its parts. RFMOs manage particular fish stocks or, at best, manage fishing effort in particular regions. This approach can be called microregulation as opposed to macroregulation, which would mean curtailing overall (global) fishing effort. A microapproach to fisheries management can of course help to protect those specific species or regions in which it is practiced, but it has no direct impact on overall or macroscale fishing capacity. Regulating specific fishing efforts in the absence of regulation of capacity has a potential negative externality or spillover effect. Capacity that is prevented by regulation from fishing for a specific species in a specific location will face a choice between remaining idle or finding something or somewhere else to fish. Remaining idle generates no personal income and yields no return on capital. Fishers and vessel owners who face RFMO regulation therefore have a significant incentive to fish for other species or in other places. Microscale regulation of particular fisheries in the absence of macroscale regulation has the negative externality of putting even more pressure on other fisheries.

We call this phenomenon the balloon problem of global fisheries regulation: squeeze fishing capacity in one place, and it simply bulges out in another. The idea of a balloon problem is similar to what others have called the problem of roving bandits in global fisheries: fishers can rove beyond a given area to escape regulatory restrictions.[36] But the balloon metaphor better captures the problem's systemic aspect. The metaphor of roving bandits suggests as a solution stronger enforcement of local property rights—in other words, the bandit problem can be solved by better local policing (or by buying off the bandits). But the balloon metaphor suggests that effective policing at a local scale does nothing to reduce the overall problem at a global scale.

Ballooning is only an issue to the extent that there are gaps in global regulation so that capacity forced out of one fishery finds its way into others. But existing gaps are huge, and even the most concerted efforts to fill them are unlikely to be fully successful. The key gaps are geographic and biological; regions and species not effectively governed by national regulators or by the existing network of RFMOs. Not all high-seas and migratory species are covered by existing RFMOs. And many commercially exploited fish stocks that are to be found entirely within particular EEZs are in effect unregulated because the government in question either fails to effectively enforce its right to regulation or sells fishing rights without adequately planning for or monitoring for sustainable exploitation.

But there are gaps in regulatory content as well. Filling these gaps would require regularizing regulation across all RFMOs, creating common sets of rules and decision-making procedures that cannot be arbitraged by fishers. It would also require agreement by member countries across RFMOs to deal with all quota questions equivalently rather than letting politics interfere differentially in the setting of quotas for various species and in different regions. It would require regularizing policies between RFMOs and coastal countries as well so that fishers would not succeed in finding gaps in regulation by moving back and forth between domestic and international fisheries. This level of coordination both among RFMOs and between them and coastal states seems highly unlikely, particularly because many countries cannot manage this level of

---

36. F. Berkes, T. P. Hughes, R. S. Steneck, J. A. Wilson, D. R. Bellwood, B. Crona, C. Folke, et al., "Globalization, Roving Bandits, and Marine Resources," *Science* 311 (2006): 1557–1558.

coordination domestically. And it would also require participation by all states in all RFMOs.

In a simple economic model of supply and demand, excess fishing capacity and therefore the balloon effect should not be a long-term problem. Regulation that has the effect of limiting the activity of fishers and of fishing capital should reduce the marginal profitability of the fishery overall. But by restricting supply, regulation should also increase the price of each fish. Reduced overall profitability should drive the most marginal fishers out of the industry, increasing the proportion of the quota for those who remain. A new equilibrium would then be reached, with fewer fishers, each catching fewer but individually more valuable fish than before regulation. From this perspective, the balloon problem should be self-correcting, as regulation drives out some capacity.[37]

But political involvement in the industry means that this logic does not work in practice to eliminate excess fishing capacity. The first problem is subsidies, which undermine the logic of the market by allowing fishers and vessel owners to stay in the industry even when market forces are trying to push them out. Many governments heavily subsidize their fishing industries in a variety of ways, some more pernicious than others, as chapter 6 discusses in detail. Eliminating or at least minimizing subsidies is a necessary step in overcoming the balloon problem by encouraging fishers and vessel owners to respond to signals from the market about overcapacity.

The logic of a simple model of supply and demand also assumes that the industry will respond to decreases in income by fishing less. Unless demand for a particular fish is so inelastic that a decrease in quantity available because of a quota is more than offset by an increase in the price, yielding an increased total income as a result of a quota (an unlikely situation given the demand substitutability of individual fish species), a quota will have the effect of lowering the total value of the particular fish species that it covers.[38] As individual fishers' income is pushed down by

---

37. Suzanne Iudicello, Michael L. Weber, and Robert Wieland, *Fish, Markets, and Fishermen: The Economics of Overfishing* (Washington, DC: Island Press, 1999).

38. Rögnvaldur Hannesson, "Effects of Liberalizing Trade in Fish, Fishing Services, and Investment in Fishing Vessels," paper prepared for the OECD Committee on Fisheries, 87th Session, March 2001, http://www.oecd.org/dataoecd/1/11/1917250.pdf.

the catch limits imposed by a quota, the most marginal fishers or boats should leave the fishery. This departure will leave more fish available for each of the remaining fishers or vessels. This process should continue until enough fishers or vessels have left the fishery that those remaining can operate at a profit despite the quota.

Because marginal fishers have both human and financial capital invested in the fishing industry, however, they are more likely to move on to other fisheries when they exit fisheries in response to quotas or other regulations—this movement is the mechanism of the balloon problem. When they cannot move on, though, a simple model of supply and demand predicts that they will leave the industry altogether rather than continue to operate at a loss. This model is predicated on the assumption (among others) of an upward-sloping supply curve, where an increase in the price of a good will lead to more being supplied by the industry, and a decrease in price will lead to less. But not all supply curves are upward sloping. For example, research on the supply of labor suggests that at points it is backward bending.[39] At certain income levels, as the price offered for labor decreases, laborers work more rather than less because they are concerned with maintaining a certain income level. Figure 3.1 provides an image of how such a curve looks. The operative question here is whether the supply curve for fisheries is like the supply curve for labor.

There are reasons to think that it can be. Fishing as an industry has barriers to exit, meaning that at the margins there are reasons for fishers and owners of fishing capital to stay in the industry and fish more rather than leave the industry in response to lower prices. These barriers are both financial and cultural. The primary financial barrier is the threat of losing a boat. Maintaining fishing vessels involves some fixed costs. To the extent that vessels are debt financed, the largest part of these costs may be fixed payments on the debt. Failure to meet fixed costs may mean loss of the vessel as well as of such capital, both financial and human, invested in it. When faced with revenue that decreases past the point of being able to cover fixed costs, owners face a choice between marginally increasing fishing effort to cover costs or losing their existing capital. As long as the marginal revenue of fishing is greater than the marginal cost,

---

39. John Pencaval, "Labor Supply of Men: A Survey," in *The Handbook of Labor Economics*, ed. Orley Ashenfelter and Richard Layard, 3–102 (New York: Elsevier, 1986).

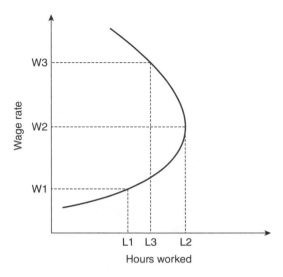

Figure 3.1
A Backwards-Bending Supply Curve for Labor.

owners are likely to choose increased effort over exit from the industry and the concomitant loss of capital.[40]

Even if individual fishers do not have financial capital committed to the industry, they often have human capital committed. Their skill as fishers commands a wage premium that they would be unable to command in other industries. If they can leave the industry temporarily in response to lower income and know that they can come back, they may well leave. But to the extent that there are barriers to getting back into the industry once they have left, they are less likely to leave in the first place. Barriers to reentry ironically may include permit systems designed to limit the number of fishers. If they know that when they leave and lose their permit, they will not be able to come back, they are less likely to leave in the first place (and forego their comparative advantage in skills). If regulations to limit the number of people fishing are poorly designed,

---

40. See, for example, Michael Basch, Julio Pena, and Hugo Dufey, "Economies of Scale and Stock Dependence in Pelagic Harvesting: The Case of Northern Chile," *Cuadernos de Economia* (Santiago, Chile) 36 (1999): 841–873. For a survey of the literature on this topic, see Linda Nøstbakken, Olivier Thebaud, and Lars-Christian Sørensen, "Investment Behavior and Capacity Adjustment in Fisheries: A Survey of the Literature," *Marine Resource Economics* 26 (2) (2011): 95–118.

therefore, they can ironically become barriers to exit from the industry. Thus, although in the aggregate it would make sense for the most marginal fishers to leave the industry in response to quotas, for a variety of reasons it may not necessarily make economic sense for individual fishers or vessel owners to do so.[41]

Barriers to exit can be cultural as well as financial. Fishing is not just an economic activity. It is often a cultural legacy, practiced across generations and in communities that are organized around fishing. Leaving the industry can thus mean not only leaving a way of life but also undermining the viability of that way of life by weakening the community upon which it is based. This cultural component of parts of the global fishing community presents one of the most intractable challenges to sustainable global fisheries governance. Policies to limit catches are in direct tension with policies to support fishing communities (by things such as subsidies) as technology improves and capital increases (meaning fewer fishers are needed to catch a given amount) and are a significant element of the balloon problem (an argument developed in more detail in chapter 6).

Supply curves in global fisheries will not always be backward bending, but there is evidence that they are so at least some of the time.[42] And as long as they are ever backward bending, they will contribute to a balloon problem in the absence of regulatory efforts to reduce fishing capacity. Technological improvements exacerbate this problem[43] because they mean that, in a world in which fish resources are static at best, participation in global fisheries needs to be constantly decreased in order to prevent such improvements from increasing capacity beyond what stocks can sustain.

As well as skewing fishers' response to regulation and therefore the level of fishing effort, backward-bending supply curves for fisheries can

---

41. On the dangers of poorly designed regulations, see Clark, *The Worldwide Crisis in Fisheries*.

42. J. Samuel Barkin and Kashif Mansori, "Backwards Boycotts: Demand Management and Fishery Conservation," *Global Environmental Politics* 1 (2) (2001): 30–41.

43. Rögnvaldur Hannesson, "Growth Accounting in a Fishery," *Journal of Environmental Economics and Management* 53 (2007): 364–376; James E. Kirkley, Chris Reid, and Dale Squires, "Productivity Measurement in Fisheries with Undesirable Outputs: A Parametric, Non-stochastic Approach," unpublished manuscript, 12 January 2010.

also skew fishers' responses to demand stimuli in the opposite way than is intended. Some attempts at protecting the world's marine living resources work through demand management rather than supply management by reducing the amount of a species demanded by the market. Consumer boycotts of particular species (as sometimes called for by the sustainable fisheries movement) or the development of aquaculture as a substitute for the fish caught in the wild are intended to reduce the pressure on global fish stocks.[44] Because a decrease in demand reduces the price of fish at any given quantity, it can result in fishers catching more of the wild stock in order to meet costs. An example of this pattern with respect to aquaculture can be found in the Pacific salmon fishery in the 1990s. Large increases in the quantity of farmed salmon forced down the price of wild Pacific salmon, leading to increased effort by fishers trying to meet costs and decreased effectiveness of binational US–Canadian efforts at sustainable management of the fishery.[45] Attempts to protect fish stocks by reducing demand for those stocks can thus both ironically and counterintuitively have the opposite effect.

Do Fisheries Balloons Happen?

A key factor in determining when fishing regulations are most likely to generate a balloon response is substitutability. Substitutability is the extent to which a particular group of fishers, given their capital stock (the type, size, and range of the vessel, the kind of equipment installed), can find new fish stocks to catch instead of those stocks that are newly regulated. Fishers who have low substitutability are often most interested in effective fisheries regulation and most willing to pressure their governments for such regulation because they know that they are stuck with their particular fishery. If it is overfished, they have few alternatives. Conversely, fishers with a higher degree of substitutability are often less interested in effective regulation. They know they can benefit in the short term by overfishing and then simply moving on to another stock in the medium term.

---

44. Barkin and Mansori, "Backwards Boycotts."

45. Samuel Barkin, "The Pacific Salmon Dispute and Canada–US Environmental Relations," in *Bilateral Ecopolitics: Continuity and Change in Canadian–American Environmental Relations*, ed. Philippe Le Prestre and Peter Stoett, 197–210 (London: Ashgate, 2006).

Conflicts over fisheries management are at their most severe when particular international fisheries are exploited by different national fleets with varying levels of substitutability.[46] For example, the independence of Namibia in 1990 created a conflict between Namibian fishers and the Spanish fleet, which had been fishing in Namibian waters. Namibia declared an EEZ and attempted to remove foreign fishers. The resource mattered most to the Namibian fishers because they had no ability to fish elsewhere, whereas Spain has one of the major global distance-fishing fleets, which made Spanish fishers much less concerned about the long-term health of the fish stocks in Namibian waters because they could simply fish elsewhere if the stocks were depleted.[47] The most substitutable fleets are at the core of the balloon problem. Not only are they the most capable of moving on to new fisheries when excluded from existing ones, but they are also the most likely to lobby against stringent fishery regulation in the first place.

Is there evidence that the balloon phenomenon actually happens? Fishers have historically always moved to follow the available fish stocks, whether their inability to access fish came about because of regulatory activity or because of simple stock depletion from overfishing.[48] It can be surprisingly difficult to trace systematically individual fishing vessels' behavior in response to regulatory action, but there are several ways to discern fisheries balloons in action. The first way is the big picture. Over time, both on the high seas and in domestic waters, catch quotas have become increasingly strict. Yet the number of fishing vessels and the overall catch have remained constant or even risen.

As figure 3.2 demonstrates, global marine fish catches increased dramatically during the second half of the twentieth century but have held relatively steady since the mid-1990s. RFMO quotas and domestic EEZ fishing regulations have become increasingly strict over that period of time. And yet overall fish catches have not declined, suggesting that

46. Barkin and DeSombre, "Unilateralism and Multilateralism."

47. "Namibian Court Orders Seizure of Five Spanish Fishing Vessels," Radio Nacional de España, Madrid, *BBC Summary of World Broadcasts*, 12 April 1991, at Lexis/Nexis.

48. Daniel Pauly, Reg Watson, and Jackie Adler, "Global Trends in World Fisheries: Impacts on Marine Ecosystems and Food Security," *Philosophical Transactions of the Royal Society, Biological Sciences* 360 (2005): 5–12.

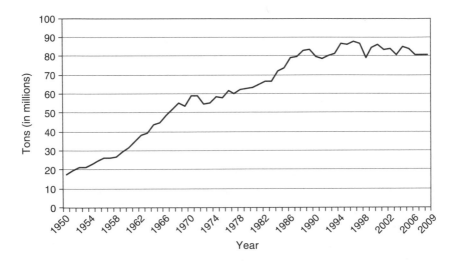

Figure 3.2

World Marine Fish Catches, 1950–2007. *Source:* Statistics from FAO, Fisheries and Aquaculture Department, "Fishery Statistics: Global Capture Production, 1950–2008," at http://www.fao.org/fishery/statistics/en.

fishing effort has moved to new locations when stocks in certain areas became regulatorily unavailable.

There is other evidence that total global fishing effort has not decreased during this period of increasing regulation. A study of global fishing effort between 1950 and 2010 concluded that there has been in increase of 1.1 percent annually since 1960, a trend that is unchanged by periods of increased global and domestic regulation.[49] The registered tonnage of all industrial-size categories of fishing vessels (the ones capable of changing regions or species to pursue new fishing opportunities when fishing in certain regions or for particular species is restricted) has similarly increased since 1970.[50] So if there are as many or more ships exerting as much or more fishing effort and catching as many fish over the past twenty years of RFMO regulatory restrictions, there are fishing balloons.

There is evidence of individual fishers' or fishing fleets' movement in response to regulatory efforts in a fishery. One recent study traced

49. J. A. Anticamara, R. Watson, A. Gelchu, and D. Pauly, "Global Fishing Effort (1950–2010): Trends, Gaps, and Implications," *Fisheries Research* 107 (2011): 131–136.

50. Ibid., 133, 134.

behavior in the Irish cod fishery in response to regulation of cod catches within a particular area. Although fishing effort decreased overall (in conjunction with vessel buyback schemes), fishing by the fleet also increased outside the regulatory area, and catches of pollack, hake, and other species increased as catches of cod declined. Fishers also changed the types of gear used.[51]

Governments themselves sometimes direct fishing effort elsewhere as an intentional policy effort. Japan, for instance, reduced its trawl fishing in the East China Sea in the 1950s by converting trawlers into salmon catchers and diverting holders of licenses from that area into the tuna longline fishery instead.[52]

Finally, it is possible to trace the movement of a given state's fishing vessels over time. Spain is an obvious state to track; its fleet is almost entirely a distance fleet, and it is the largest fishing fleet in Europe.[53] In the mid-1980s, it found itself closed out of European waters for ten years as a condition of Spain's accession to the EU.[54] After a period fishing for hake in Namibian waters following EU accession, Spanish vessels moved to the Northwest Atlantic to fish for a variety of turbot species in the mid-1990s.[55]

Figure 3.3 represents catches of all Spanish-registered fishing vessels during the period from 1985 to 2010 for three different species (Atlantic cod, Greenland halibut or turbot, and skipjack tuna) in two different regions (the northwestern Atlantic and the eastern central Pacific). As each of these first species were regulated, overall Spanish catches moved to other areas. It is worth noting that this phenomenon does not even account for vessels owned or crewed by Spanish citizens that are re-registered in FOC registries in order to get around existing rules that bind Spanish vessels.

51. Sarah Davie and Colm Lordan, "Examining Changes in Irish Fishing Practices in Response to the Cod Long-Term Plan," *ICES Journal of Marine Science* 68 (2011): 1638–1646.

52. H. Kasahara, "Japanese Distant-Water Fisheries: A Review," *Fishery Bulletin* 70 (1972): 227–282.

53. Peter Gruner, "The New Spanish Pirates," *Evening Standard*, 13 April 1995.

54. Paul Koring and Kevin Cox, "Scots Support Seizure of Vessel," *Globe and Mail*, 11 March 1995.

55. Barkin and DeSombre, "Unilateralism and Multilateralism."

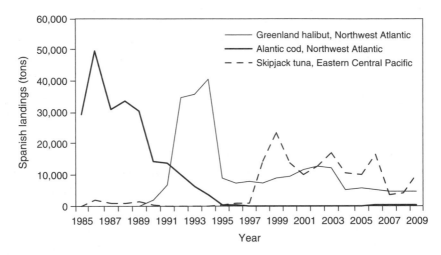

Figure 3.3

Movements of the Spanish Fleet, 1985–2009. Statistics from FAO, Fisheries and Aquaculture Department, "Fishery Statistics: Spain Landings by Species, 1985–2009," http://www.fao.org/fishery/statistics/en.

It thus seems clear empirically that both in the aggregate and in the response by specific fleets to fishing regulations for certain species in particular areas, fishing vessels are continuing their fishing efforts for different species or in different locations. The existence of these balloons suggests that regulation by states within their waters or by RFMOs for particular species in a given region cannot successfully manage global fisheries.

Conclusions

Microregulation of global fisheries cannot provide adequate systemic management of global fisheries. This conclusion is overdetermined—there is a laundry list of reasons why current regulatory patterns do not on their own suffice to regulate fisheries on a sustainable basis. And the evidence certainly seems to bear out the claim that current patterns of microscale regulation are less than ideally successful. This chapter has laid out some of these reasons. Microscale regulation on its own will ultimately fail in providing effective governance of global fisheries because it is based on goals that are unclear, on analytic methods that are inadequate, and on a model of regulation that tends to displace rather than solve the core problem of overfishing.

Unclear goals are manifested most clearly in the unresolved tension between analysis based on environmental economics and CPRs and analysis based on population biology and MSY. Neither set of analytical tools is individually up to the demands placed upon it. Fisheries management models are premised on information that is both more complete and more tractable than is information about the marine ecologies we exploit for food. Population dynamics within species and ecological relationships among species are too complex and involve too many scientific unknowns to model with the accuracy necessary to manage fisheries through precise quotas and other species-specific management tools. Furthermore, the interactions between the two leave regulatory and political gaps that encourage gaming the system through subsidization and regulatory arbitrage, so that even when an RFMO gets microregulation right, the regulation still does not scale up into an aggregate pattern of sound macromanagement because of the balloon problem.

Management of fishing capacity in conjunction with regulation of catch levels of specific fish stocks might help address these shortcomings. It would most clearly address the balloon problem. This problem is specifically one of capacity, and macrolevel management in this case is specifically about cutting capacity. If overall capacity is limited to what the global marine ecosystem can bear, then limiting access to a particular stock to what that stock can bear should not create unbearable pressure elsewhere. If catch regulation is driven by decreasing stock health, then it needs to be accompanied by a decrease in overall capacity. Capacity restrictions would not need to be adjusted with every new quota decision. Because capacity restrictions would be based on what global fisheries overall can bear, general adjustments reflecting the overall health of marine ecosystems should alleviate the balloon problem.

Macrolevel regulation would also address the information problem faced in the use of the standard analytical tools for fisheries management. It would not help, it is true, with the creation or modeling of information about fish stocks. But it would make the politics of management much less sensitive to modeling precision. Microregulation based on fisheries models and on concepts such as MSY is highly sensitive to political interpretation of data. To the extent that microregulation represents the only form of restriction on fishing effort, fishers have an incentive to push back against regulation in all cases. They may not succeed in all cases, but they are likely to succeed in some. If macroregulation restricts capacity, however, it will undermine the incentive for fishers to push back politically against microregulation. Because a reduction in overall overcapacity reduces the extent to which microregulation creates species- and region-specific overcapacity, such capacity reduction would reduce the short-term negative economic consequences of microregulation on specific fishers and thereby reduce the incentive to engage in political activity to weaken microregulation. This political effect should be more pronounced the greater the extent to which fishers can move their fishing effort across species or fishing areas. As such, macroregulation is likely to be more effective at changing the politics of international fisheries management as fishers become more able to substitute across species and regions.

Given these arguments in favor of macromanagement, why has international fisheries politics focused predominantly on micromanagement? One reason is the structure of international fisheries governance, as explained in chapter 2. This governance happens primarily through RFMOs, which in turn are subject to member states' political inputs. But as chapter 4 suggests, this political input comes from a specific subpart of states. Although in principle member states of RFMOs are members as states, in practice fisheries bureaucracies within states generally take the active role in participating in RFMO politics and governance. And the practical constituencies of these fisheries bureaucracies are not populations at large, with concerns for aggregate environmental management or even economic maximization. The primary constituents of fisheries bureaucracies are for the most part fishers with an interest in maximizing the well-being of the fishing industry and its current members. Chapter 4 develops these arguments and explains the effect of these structures on fisheries management at the global level.

4

Regulatory Capture

Global fisheries governance suffers from a problem of regulatory capture, a situation where the entity entrusted with overseeing the public interest instead acts in the interests of industry, as though "captured" by it.[1] This capture is primarily a problem at the national level, but because international fisheries regulation is in effect managed cooperatively by national regulators, the problem re-creates itself at the international level. This pattern favors microregulation over macroregulation because actual capacity reduction would require reducing the size of regulators' constituencies, but this regulatory capture also undermines the effectiveness of micromanagement.

Most national fisheries regulators are industry regulators rather than environmental regulators. They tend to have a bureaucratic prominence that is disproportionate to the economic importance of fisheries to national economies. Furthermore, fishing is one of the few industries in which environmental effects are regulated by industry regulators, as this chapter demonstrates, rather than by environmental regulators. Thus, despite marine ecosystems' environmental sensitivity, there is reason to expect that fisheries are less well regulated than other ecosystems because of the absence of the bureaucratic check of an environment bureaucracy. This problem is exacerbated at the international level, when national fisheries bureaucracies are explicitly tasked with maximizing the interests of the national industry and of the national budget. Even if these regulators have a personal bias toward sustainable management, they face a structural conflict of interest.

---

1. M. E. Levine, "Regulatory Capture, Public Interest, and the Public Agenda," *Journal of Law, Economics, and Organization* 6 (1990): 167–198; Jean-Jacques Laffont and Jean Tirole, "The Politics of Government Decision-Making: A Theory of Regulatory Capture," *Quarterly Journal of Economics* 106 (4) (1991): 1089–1127.

In addition, members of the fishing industry (rather than or in addition to national regulators of that industry) participate in management as national representatives to RFMOs and other international bodies involved directly in the creation of the rules of international fisheries management. These instances constitute an almost literal case of regulatory capture. But even when regulators are ostensibly independent from industry, some level of regulatory capture in fisheries management is likely.

One might respond that fishers themselves, the constituencies that fisheries bureaucracies represent, should have an interest in sustainable management. After all, a successfully managed fishery will provide employment opportunities indefinitely. But this interest does not do as much as it could to guide actual management decisions, for several reasons. Fishers, like all economic actors, discount the future somewhat. Even if given clear property rights over fish stocks, they will not always choose sustainable management when there are advantages to be gained from short-term exploitation. Second, the same bureaucracies that regulate fishers also for the most part subsidize them, creating a complicated set of political dynamics between industry and regulator. Finally, even when fishers have a preference for sustainable management, there are still distributional issues in fisheries management, and there is no optimal way of dealing with these issues. The politics of distribution will always plague efforts at the rational management of fisheries resources.

This chapter develops these arguments and provides examples of patterns of regulatory capture in both national and international fisheries governance. Its argument is that macroregulation of global fishing capacity is unlikely as long as the key participants in governance serve the interests of the industry rather than of sustainable management. Effective macroregulation requires either that a critical mass of national fisheries regulators distance themselves effectively from the industry's immediate concerns or that the institutional focus of macroregulation move away from RFMOs and their national regulator members and toward other institutions with different national constituencies.

National Fisheries Regulation

Although the fishing industry is small in most countries relative to the size of the overall economy, the governance of fishing often has an undue bureaucratic prominence. This industry is in some instances one of very few (along with agriculture and finance) whose regulators have representation at the cabinet level in national governments. It is also arguably the only resource-extraction industry for which as a general rule there is no regulatory oversight by separate environmental regulators who are bureaucratically distinct from the industry regulator. Furthermore, the core constituency of fisheries regulators is often unclear. These regulators are generally asked, as a central element of their mission, to manage national fishery resources and to contribute to the management of international fishery resources with an eye to both sustainability and maximum yield.

But regulators are often also called upon to help maintain the socioeconomic character of existing commercial fishing communities and cultures, a goal that is not necessarily compatible with resource sustainability. For example, the fisheries regulator in Peru, home of the world's second-largest catch in years with good anchovy runs, is tasked "to balance the sustainable use of aquatic living resources, conservation of the environment and socio-economic development."[2] Furthermore, fisheries regulators are often participants in government efforts to use fishery resources either as a vehicle for industrial development or as a way to extract rent in the form of license fees from either the domestic industry or from foreign governments. The potential conflicts among these demands on fisheries regulators serve to undermine ecologically effective fisheries management.

Some countries have cabinet-level departments or ministries devoted exclusively or primarily to fisheries. Norway, for example, has a Ministry of Fisheries and Coastal Affairs; Canada has a Ministry of Fisheries and Oceans; Malaysia has a Ministry of Marine Affairs and Fisheries; and Vietnam has a Ministry of Fisheries. In the EU, which is an important actor in global fisheries regulation, the Directorate-General for Maritime Affairs and Fisheries, the equivalent of a cabinet-level department, is run by a commissioner, the EU equivalent of a cabinet minister.

A more common bureaucratic structure is for fisheries regulation to be a second-tier bureaucracy within a cabinet-level department or ministry. In most European countries with large fishing industries, for example, the fisheries regulator is housed within the Ministry of Agriculture or its

---

2. FAO, *National Fisheries Sector Overview: Peru*, FID/CP/PER (Rome: FAO, May 2010), 9, ftp://ftp.fao.org/FI/DOCUMENT/fcp/en/FI_CP_PE.pdf.

equivalent (a significant exception is Spain, where a recent reorganization has left it as part of the Ministry of Environment and Rural and Marine Affairs). In Peru, it is part of the Ministry of Production and Economy, and in Chile it is part of the Ministry of Development and Reconstruction. In some cases, fisheries regulation shares a cabinet-level designation with agriculture, as in Japan, Russia, and Bangladesh (in the latter case, the Ministry of Fisheries and Livestock).

Only rarely are fisheries regulators to be found in third-tier bureaucracies in countries with substantial fishing industries (examples include the United States, where the Fisheries Service is part of the National Oceanic and Atmospheric Administration, itself part of the Department of Commerce, and India, where fisheries are managed from within the Department of Animal Husbandry and Dairying in the Ministry of Agriculture).

Complicating the picture of fisheries regulation is a frequent problem of multilevel governance. This problem is most straightforward at the international level, where regulation of high-seas fisheries can only be accomplished cooperatively rather than authoritatively. But significant problems exist in federal and supranational systems as well. In federal systems such as the United States and Canada, individual states and provinces have rights over some aspects of fisheries governance, but not all. In the United States, for example, individual states have regulatory authority out to three miles, but the federal government has authority in the EEZ. Furthermore, governments at the state/provincial level can become political proponents and subsidizers of local fishing industries, knowing that the political costs of effective fisheries regulation will likely be born by the federal government. This phenomenon is known in the rational choice literature as "cheap talk,"[3] so called because lower levels of government can express political support for the industry and hope to get its support in turn, knowing that difficult but necessary management decisions are the responsibility of a different level of government.

The EU case in this context is similar to that of a federal system, but one with a particularly weak central government. The EU, through the

---

3. Jack L. Goldsmith and Eric A. Posner, *Moral and Legal Rhetoric in International Relations: A Rational Choice Perspective*, John M. Olin Law and Economics Working Paper Series no. 108 (Chicago: University of Chicago Law School, 2000); Thomas Risse, "International Norms and Domestic Change: Arguing and Communicative Behavior in the Human Rights Area," *Politics & Society* 27 (4) (1999): 529–559.

European Commission, has the authority to regulate fisheries resources in EU waters, but national-level governments often promote and subsidize their national fishing industries in competition with other EU members' industries. The EU is, however, weaker as a fisheries regulator than most federal governments. Although the commission has the authority to regulate fish catches and impose quotas, it does not have any independent capacity to monitor and enforce its rules. It counts on national governments to do so with respect to their own fishers, which creates a classic free-rider problem and the kind of two-level enforcement game discussed in chapter 3. In addition, national governments have a much stronger direct voice in EU-level governance than do most lower levels of government in federal systems. Thus, the least responsible governments from a fisheries management perspective, such as Spain, have a greater ability to affect EU-level decisions and to undermine effective management than do lower-level governments in federal systems. The effect of this two-level process is sufficiently pronounced that the EU fisheries commissioner recently characterized the common fisheries policy as a "disaster."[4]

These problems of multilevel governance contribute to lighter regulation than would otherwise be the case and to greater levels of subsidization. The problem of regulatory levels is exacerbated by the frequent absence of outside environmental oversight of fisheries regulators. Industry regulators control rights to and levels of resource extraction. To the extent that there is any involvement of environmental bureaucracies in oversight of fishing practices, it generally focuses not on levels of extraction per se, but on the negative externalities of extraction, in particular pollution. Pollution from fishing vessels is indeed often regulated either by environmental regulators or by international treaty. But this regulation does not intersect with the problem of overfishing, leaving the issue of overfishing to be overseen only by the industry regulator.

The absence of outside environmental oversight is a problem in this case because of the multiple demands placed on fisheries regulators. When an industry regulator shares oversight responsibility with an environmental regulator, the latter's bureaucratic focus can generally be assumed to be the control of environmental externalities rather than the health of the

---

4. EU fisheries commissioner Joe Borg, quoted in "Charlemagne: A Commission Report-Card," *The Economist* (26 September 2009), 68.

industry in question (although industry capture of environmental bureau-cracies can be a problem as well). For example, in the United States the Environmental Protection Agency would be expected to evaluate the en-vironmental effects of a new coal mine before approving it, but we would not expect it to be arguing for the subsidization of the coal industry at the same time. That is precisely what happens in most fisheries bureaucracies. Outcomes, in terms of the trade-off between the industry's environmental impact and the economic cost to the industry of minimizing that impact, can be understood as the result of political and bureaucratic competition between two regulators, each of which has a clear and distinct priority. But in the case of fisheries regulators, there are often no clear priorities within the one regulator whose focus is the level and type of fishing ac-tivity and no external bureaucracy to focus on the direct environmental impacts of that fishing activity (rather than its externalities). And in cases where the priorities are clear, environmental sustainability is likely not to be the focus.

International Fisheries Regulation

The same national fisheries regulators that are implicated in the subsidiza-tion of excess capacity domestically are the voting participants in interna-tional fishery regulation. As a result, the same sort of regulatory capture seen at the domestic level can be seen within RFMOs as well. As discussed in chapter 2, these organizations have a multistage decision-making pro-cess, and industry's short-term interests can trump the protection of the resource at any of the stages. First, a scientific committee recommends a sustainable catch level. Second, the commission itself votes (usually by some kind of supermajority rather than by unanimity) to pass rules about catch levels or any other restrictions. Finally, in commissions where votes need not be unanimous, states that are not happy with the resulting rules are able to "object" or opt out of the rules. This structure augments the already existing opportunities for regulatory capture to decrease the level of regulatory oversight.

The system of regulation within RFMOs relies on the idea that scien-tific information about the health of stocks can be translated into politi-cal decisions to restrain fishing behavior to that which can be supported by existing stock levels. The scientific committee is staffed, albeit with exception, by those who are focused on the fish stocks' actual health, although the scientists' background and experience varies considerably.[5] Regulatory capture is not likely at this stage, although to the extent that industry-related scientists become involved with the process, it is not impossible. The commission, however, is composed of representatives of states, and both the states and the commissioners have political agendas.

Fishing regulators, at the international as well as the domestic level, are not generally from environment or conservation backgrounds. Although commissioners are charged with representing their states, many individual commissioners have direct ties to the fishing industry, the most direct form of regulatory capture. At least one member of a state's delegation to an RMFO usually comes from the state's fisheries agency (which, as discussed earlier, is not usually an environmentally oriented agency). If there are other members, they frequently come directly from some aspect of the fishing industry. For instance, the Indonesian delegation to the CCSBT meetings includes the chair of the Indonesian Tuna Association, the managing director of Sekol Farmed Tuna, the secretary-general of the Indonesian Tuna Longline Association, and the president of the Harini Group (a fishing labor-recruitment company), along with two fisheries ministers and a research scientist.[6] Four of the eight EU commissioners to the Western and Central Pacific Fisheries Commission (WCPFC) are heads of industry groups.[7] These examples are representative of the composition of delegations to those commissions.

Other commissions by design include representation from the different aspects of the fishing industry: the International Pacific Halibut Commission (an organization for cooperation between the United States and Canada) points out that its tradition has been that "one commissioner from each country has been an employee of the federal fisheries agency,

---

5. Tom Polachek, "Politics and Independent Scientific Advice in RFMO Processes: A Case Study of Crossing Boundaries," *Marine Policy* 36 (2012): 132–141.

6. CCSBT, *Report of the 15th Annual Meeting of the Commission (14–17 October 2008)* (Deakin, Australia: CCSBT, 2008), appendix 2, http://www.ccsbt.org/userfiles/file/docs_english/meetings/meeting_reports/ccsbt_15/report_of_CCSBT15.pdf.

7. WCPFC, "Commission for the Conservation and Management of Highly-Migratory Fish Stocks in the Western and Central Pacific Ocean, Fifth Regular Session, Busan, Korea, 8–12 December 2008," 2009, p. 47, http://www.wcpfc.int/node/1892.

one a fisher, and one a buyer or processor."[8] In that case, the commission's composition has an especially important potential impact on management because measures can be passed by a two-thirds vote of the commissioners from each state.[9] At least theoretically, then, the industry-based commissioners can outvote the commissioners from the fisheries agencies.

In their capacity as members of RFMOs, national regulators have a dual role. They are both participants in collective management and partisans for their national industries. The fact that it is national regulators that are participants in the diplomatic process of international fisheries management, rather than national governments as rational unitary actors, can have a number of effects. For example, it creates a sensitivity not only to the industry's demands for quota, but also to the political effects of smaller quotas and to regulators' ability to attract subsidies from general government funds to compensate for smaller quotas.

Another expected effect of national regulators' dual role is RFMO's reluctance to constrain members' ability to fish. There is a clear collective-action problem: each regulator wants a bigger quota for its own nationals even though at the same time long-term management requires catch limits to decrease. This problem is exacerbated by the fact that even when specific national representatives are regulators rather than industry actors, they come from a fisheries bureaucracy rather than having a specifically environmental mandate. High discount rates and concerns about free riders can also make regulators less willing to pass catch limits that are as strict as than the current state of the resource requires. The presence of objections procedures, whereby states can opt out of catch rules that are more restrictive than they prefer, may also temper the stringency of rules because a strict rule that few states agree to abide by may be less useful than a laxer one that everyone implements.

Regulatory capture can be most clearly seen in RFMOs in the difference between what the scientific committees recommend and what rules the commissions themselves pass. In theory, RFMO commissions should make decisions based on scientific advice about the level of fishing that can be sustained; the Straddling Stocks Agreement reaffirms this principle

---

8. International Pacific Halibut Commission, "Commissioners," n.d., http://www.iphc.int/about-iphc/27.html.

9. Convention for the Preservation of the Halibut Fishery of the Northern Pacific Ocean and Bering Sea, Article III(1).

legally.[10] Commissions, however, do not always follow the advice laid out by their scientific committees. Although they occasionally pass rules stricter than scientific advice suggests (representing precautionary management), the TAC passed by the political process more frequently exceeds the advice given. The history of scientific recommendations versus TAC decisions on shrimp catches in NAFO provides an example of this disparity (figure 4.1). Although NAFO's decisions on catch levels bear some relationship to the scientific advice, it sets TACs consistently above the advice.

It is useful to examine the relationship between scientific recommendations and commission TAC decisions across a range of RFMOs and protected species and years. We calculated the ratio of TAC to scientific advice across the species regulations within five RFMOs (CCAMLR, NAFO, NEAFC, the Southeast Atlantic Fisheries Organization [SEAFO], and ICCAT) in the recent stretch (between 1998 and 2011) for which both advice and catch limit information is publically available. We chose these five commissions because they passed substantive regulations and made available information on both scientific advice and catch levels. Calculating this relationship as a ratio means that if the number is 1, the TAC is set exactly at the scientific committee recommendation. If the ratio is lower than 1, the catch limits are set more conservatively than the scientific recommendation, but if it is higher than 1, the TAC exceeds scientific advice.

As figure 4.2 suggests, catch limits are often set higher than—sometimes dramatically higher than—the scientific advice. Moreover, the propensity for the catch limits to exceed scientific advice increases in the later years in our data set, suggesting that the trend is not in the right direction. The political process results in states collectively deciding on fishing rules that are in the aggregate noticeably less stringent than scientific advice suggests. And this observation does not even encompass the possibility that actual catches might be higher than catch regulations mandate.

Although industry capture of regulators results in higher quotas than are sustainable, there are other international regulatory approaches on which RFMO member states can agree. One such effort is to exclude nonmembers, which in this case means fishing vessels flagged in nonmember states, from fishing for regulated species or in regulated areas. The

---

10. Straddling Stocks Agreement (1995), Article 5(b).

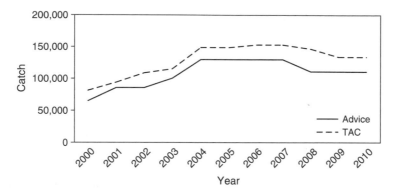

Figure 4.1

NAFO Shrimp Scientific Advice on Limits Compared to TAC, 2000–2010. *Source:*
NAFO annual Scientific Council reports and annual reports for the years covered,
available at http://www.nafo.int.

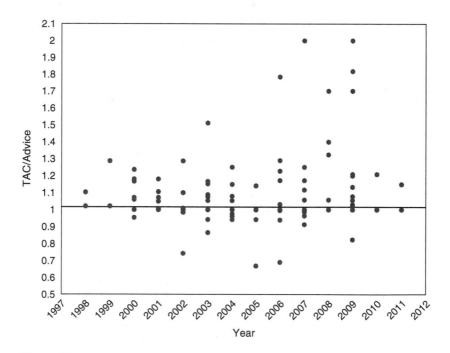

Figure 4.2

Regulator Catch Limits Compared to Scientific Advice on Limits, 1997–2012.
Source: Calculated from information presented in annual reports and scientific
committee reports for the relevant RFMOs (CCALMR, NAFO, NEAFO, SEAFO,
ICCAT). When scientific committee advice was given as a range, the midpoint of
the range was used.

demands of industry support and sustainable environmental management are congruent in these efforts—both are improved by keeping vessels that are not regulated by the system from fishing. And although such efforts are particularly difficult in the setting of international law, RFMOs and their members have made some real progress.

In principle, the 1995 UNCLOS Straddling Stocks Agreement requires that all states that participate in an international fishery participate in collective management as well, although there is no enforcement mechanism for this requirement. In practice, mechanisms of exclusion include preventing vessels flagged in nonmember states from landing their catches in member state ports and keeping fish caught outside of international regulatory processes from member states' markets. ICCAT pioneered a system of trade restrictions against states whose vessels regularly fished in the ICCAT regulatory area without following ICCAT rules. CCAMLR created a catch documentation system under which CCAMLR member states are not allowed to land or transship fish that do not have the documentation to demonstrate that they have been caught within the regulatory process.[11]

RFMOs do some important work, especially when the interests of both the regulators and the fishing industry happen to align with ecological sustainability. Excluding nonparticipants is necessary for effective fisheries management, and although quotas are often far higher than sustainable management demands, they are also often (although by no means always) better than nothing. Many RFMOs are also engaged in efforts to improve their regulatory structures and management systems by implementing ecosystemic approaches to management or by engaging in adaptive management systems that implement lessons learned across different RFMOs.[12] RFMO secretariats are likely to be less affected by industry capture than the voting members are, leaving considerable scope for innovation in management and regulatory process up to the point where innovation is sufficiently threatening to industry demands in voting members that they block implementation or opt out of rules. Nonetheless, the

---

11. Elizabeth R. DeSombre, "Fishing under Flags of Convenience: Using Market Power to Increase Participation in International Regulation," *Global Environmental Politics* 5 (4) (2005): 73–94.

12. D. G. Webster, *Adaptive Governance: The Dynamics of Atlantic Fisheries Management* (Cambridge, MA: MIT Press, 2009).

RFMO system as a whole is sufficiently affected by industry capture that it cannot serve reliably as the primary basis for effective sustainable management of global fisheries.

Management Priorities

Regulatory capture of fisheries bureaucracies at both the national and the international levels is likely to affect management priorities. These priorities, both statutory and practical/political, may include ecological sustainability but are more likely to include economic sustainability. In either case, an emphasis on sustainability of any kind ensures at minimum some focus on fish stocks' long-term health. But fishery bureaucracies often have other priorities as well. They may focus on the fishing industry's sustainability rather than on the fishery per se. A focus on the industry as constituted means keeping fishing communities in place, whether those communities have access to sufficient fish stocks to be economically viable or not. This focus can lead to the institutionalization of patterns of subsidy as government is gradually called upon to increase subsidization when the fishery gradually becomes less economically viable (see chapter 6 on subsidies). Since the industry is also the regulator's core constituency, this institutionalization of subsidies in the face of economic pressure on the industry makes bureaucratic sense in that it protects the core constituency. Given a standard pattern of increasing efficiency in an industry facing static or shrinking resource stocks, the number of people the industry can support should be expected to decrease over time. Fishing communities can thus be expected to become less economically viable over time, all else being equal. A focus on the sustainability of fishing communities rather than on the sustainability of fishing practices therefore contributes directly to the maintenance of excess capacity through a logic that makes sense bureaucratically and politically, but not ecologically.

Beyond sustainability of any kind, there are two other priorities often held either by fisheries regulators or by national governments acting in lieu of effective fisheries bureaucracies. The first is economic development. Fish are, in terms of national accounting, a free resource. Exploiting a previously unexploited or underexploited fishery can seem like a way to realize latent economic value and bring that value into the national economy. Developing fisheries therefore can seem like a good way to create economic activity and generate capital that otherwise would remain latent. As a result, when governments see fisheries as a development tool, they may not approach them sustainably. They may even use them in a way that essentially treats fish stocks as a nonrenewable resource, depleting them but increasing national output in the short term. The problems with this logic and with the use of fisheries for purposes of economic development are manifold (the subject of chapter 7), but governments sometimes look to fisheries as tools of economic development nonetheless.

The second priority that can be held by fisheries regulators or governments more broadly, in particular governments of poorer countries, is economic rent. As the term is used by economists, *rent* is "any income payment in excess of that required to induce participation in the production process."[13] In other words, rent is surplus that can be captured without diminishing the level of economic activity. The selling of fishing rights, in particular rights to fisheries that cannot be accessed by domestic fishing industries (generally because they are too far from shore or in waters that are too deep), is a form of rent. When the use of fishery resources as a development tool is unfeasible (generally because it is too capital intensive for the government in question), the use of them to generate rent and therefore government revenue can remain an option. When the revenue is in fact going to the national budget, the use of a fishery resource for rent is subject to the same problematic logic as its use for development. But rent in the form of cash payments is also more easily appropriated by individuals within the government and therefore more prone to corruption.[14]

The irony of this particular problem is that the ability to capture rent and the concomitant possibility for large-scale corruption result in large part from the creation of EEZs. EEZs were supposed to help solve the problem of overfishing in the high seas by privatizing (to some extent) the international commons, giving individual countries incentives to manage their newly acquired waters sustainably. The idea behind and the problems with this effort are discussed further in chapter 9.

---

13. Richard B. Goode, *Government Finance in Developing Countries* (Washington, DC: Brookings Institution Press, 1984), 181.

14. See, for example, Quentin Hanich and Martin Tsamenyi, "Managing Fisheries and Corruption in the Pacific Islands Region," *Marine Policy* 33 (2) (2009): 386–392.

Is Regulatory Capture a Problem?

A simple rationalist response to widespread industry capture of national fisheries regulators is that this capture should not be inherently problematic. Individual fishers have an incentive to overfish an unregulated commons, but they should have an incentive to support rational management that limits access to the commons. Because fishers understand that their long-term employability as fishers depends on the sustainability of the stocks that they fish, they should support regulatory efforts as long as they have confidence that those efforts will be successfully monitored and enforced. Even more so, fishers who gain full property rights over particular fish stocks should develop a personal interest in the sustainable long-term management of the stock because ownership internalizes the negative externalities of CPR use. Efforts by national fisheries regulators to bring international fisheries under national regulatory control should improve sustainable management, as should efforts by regulators to provide clear property rights to individual fishers.

National regulators that represent fishing industry interests should be concerned with sustainable management of fish stocks. They have the incentive to achieve such a program because the industry needs sustainable management for its long-term viability, and the regulator needs the industry to be viable in order to guarantee its own bureaucratic health. Fishers in turn have an incentive to support sustainable management that guarantees that, in exchange for sharing the short-term costs of such management, they will have a share in its long-term benefits. But in practice these apparently converging interests in sustainable management often do not produce management practices that actually conserve fisheries' long-term health, even at the national level where states have the legal authority to compel participation in management schemes. The reasons for this gap between concerns and practice are understandable.

Some of these reasons are relevant to the national level, whatever the form of potential regulation considered. Many of them are directly related to the many demands on national fishery regulators other than sustainability discussed earlier, such as economic development, community maintenance, and the maximization of economic rent. Even if states acted as unitary rational economic actors with respect to fisheries management and their utility was defined as maximizing the fishing income of the domestic industry, time horizons might lead to a level of fishing higher than MSY and much beyond an ecologically sustainable level of fishing. States are of course not rational unitary actors both because of the two-level problem noted earlier, on which different levels of government are responsible for different aspects of fisheries governance, and because state decision makers may be more concerned with electability than with rational governance. Electability in turn can depend on decisions that appeal to various constituencies, including but not limited to fishers, others involved in the fishing industry, environmentalists, and those voters (in many countries a majority of the population) who care little about fisheries issues.

Beyond the two-level problem and questions of electability and good governance, there are the various demands on fisheries regulators other than sustainability. The use of national fishery resources as a development tool, for example, can decrease concern for the fishery's sustainability. Once capacity has been developed in a protected national market, it can then compete with other fishing capacity in international waters. And if the primary development concern is to create employment in the short to medium term, then the fishery can be seen as a stage in a longer-term development plan rather than as a sustainable industry in its own right. Politicians' short time horizons may lead them to prioritize economic development according to an election cycle.[15] Beyond the dictates of individual management demands, furthermore, the multiple demands on many fisheries regulators can lead directly to suboptimal results with respect to each of the demands, not just sustainability. Multiple constituencies' demands on government may not be reconcilable, leading to outcomes that are suboptimal by any measure. To the extent that Arrow's paradox[16] holds with respect to fisheries governance, privatization of fisheries at the national level may lead to management that is no better than that offered by international cooperation to manage the commons.

The failure of regulation at the national level to deal with the tragedy of the fisheries commons can be further exacerbated by state failure, particularly in poorer developing countries. States can experience a failure either of capabilities or of intentions. A failure of capabilities happens when

---

15. Zachary A. Smith, *The Environmental Policy Paradox*, 2nd ed. (Englewood Cliffs, NJ: Prentice Hall, 1995), 46–47.

16. Kenneth J. Arrow, *Social Choice and Individual Values* (New York: Wiley, 1951).

governments are simply unable to police fishing regulations due either to a lack of administrative capacity or to a lack of the physical resources, such as coast guard vessels, necessary to monitor activities within their own waters.[17] In poor countries with large EEZs and lucrative fisheries, this inability can be a real problem, as noted in chapter 2. A failure of intentions happens when government officials act in response to individual utilities rather than to the utilities of the states that they govern or to the constituencies that they represent. In other words, corruption can be a real problem. To the extent that individuals within a state structure are interested in profiting directly from the selling of fishing rights, either to a domestic fishing industry or to foreign governments, their time horizons will be defined by their expected tenure in government. They therefore may have an incentive to sell as much as they can in a short time frame, without reference to the fishery's long-term health.

Regulatory capture should also, under the right circumstances, not be a problem from individual fishers' perspective. Fishers should support policies that protect the fish stock's long-term health, and if they have undue influence on the regulatory process, they should seek to create such policies. But there are reasons, beyond those issues discussed in chapter 3 (such as the possibility of backward-bending supply curves), why individual fishers will also *not* support or uphold policies designed to protect their long-term interests in sustainable management.

One reason is that they face a time horizons problem. To the extent that fishers discount the future, they will be willing to overfish in the present as long as the effects of that overfishing are far enough in the future. In the case of long-lived fish, such as orange roughy or Patagonian toothfish, the time-horizon problem is straightforward. These fish reproduce on a timescale that is generational from a human perspective, meaning that the people who fish for them now may well be retired by the time the fish's offspring are of sustainably harvestable age. Those individuals therefore have little incentive for sustainable management in this context.

There may be little such incentive with respect to shorter-lived species as well. The literature on the social discount rate often assumes rates

---

17. On administrative capacity as a limiting factor in international environmental cooperation, see Peter M. Haas, Robert O. Keohane, and Marc A. Levy, eds., *Institutions for the Earth: Sources of Effective International Environmental Protection* (Cambridge, MA: MIT Press, 1993).

in the 6 to 8 percent range,[18] meaning that a fish caught in five years is worth between two-thirds and three-quarters of a fish caught now (depending on which end of the range one is in), and a fish caught in a decade is worth roughly half that of a fish caught now.[19] But research extrapolating from bid prices in ITQ regulatory processes suggests that in practice fishers' discount rates can be much higher than that—as much as 33.4 percent.[20] This means that the current value to a fisher of a fish caught in five years is less than one-quarter that of a fish caught now, and the value of a fish caught in a decade is less than 6 percent. Time horizons this short can lead to significant overfishing even by rational fishers with full property rights who are fully cognizant of the extent of their overfishing.

But fishers may not be fully cognizant of the extent of their overfishing. They often have a different view of fish populations than do scientists. Fish populations, as noted in chapter 3, often do not decline in a linear and gradual way. Rather, they can crash at a specific point. The individual fisher who does not have direct experience of this pattern may see the stock as perfectly healthy, whereas a fishery scientist, having seen general models of fish populations, might see evidence of unsustainable fishing and even of an impending crash.

But fishers may also intentionally make use of regulatory capture for rent seeking, which in this context involves gaining economically by manipulating the regulatory environment rather than by engaging in economic activity such as fishing. To the extent that the regulator is captured by the industry, it may come to see its role as protecting and supporting rather than regulating the industry. Protecting the industry should include protecting fishers from themselves by preventing overfishing, but it may

---

18. Samia Atoine Azar, "Measuring the US Social Discount Rate," *Applied Financial Economics Letters* 3 (2007): 63–66; Juzhong Zhuang, Zhihong Liang, Tun Lin, and Franklin De Guzman, *Theory and Practice in the Choice of Social Discount Rate for Cost–Benefit Analysis: A Survey*, European Report on Development Working Paper Series no. 94 (Manila: Asian Development Bank, May 2007).

19. More precisely, at a 6 percent discount rate, the present value of a fish caught in five years is 75 percent of the value of a fish caught now, and that of a fish caught in ten years is 56 percent. For an 8 percent discount rate, the equivalent numbers are 68 and 46 percent.

20. Frank Asche, "Fishermen's Discount Rates in ITQ Systems," *Environmental and Resource Economics* 19 (2001): 403–410.

also include interfering with the operation of an open market to give additional economic benefits to fishers.

In some cases, regulators may be able to create a more profitable economic environment for current fishers without actually increasing strain on the resource. For example, a policy that restricts entry into the fishery or otherwise privileges those that have historically fished in the area (in the distribution of catch allocations, for instance) creates economic benefit for an existing population of fishers without necessarily causing more pressure on the fish stocks. If done well, prohibiting access by new entrants might even decrease the overall use of the resource, although in practice existing fishers in this situation are likely to increase their use of the resource if no additional policies prevent it.[21]

Another form of rent seeking, however, may involve sustaining, through the mechanism of subsidies, an industry larger than the available fishery resources can bear. Chapter 6 discusses the many forms of subsidy currently provided directly and indirectly to the fishing industry globally, usually with fisheries regulators' advocacy or at least acquiescence. To the extent that regulators are captured by the fishing industry that they are supposed to be regulating, they are likely to provide a sympathetic hearing to demands from the industry both for a congenial regulatory environment and for active subsidy. The extent to which regulators are able to subsidize depends on their bureaucratic resources and their ability to draw resources from the state more broadly. To the extent that regulators are able to draw these resources from the state, fishers have an incentive to engage in rent-seeking behavior, which increases their income beyond what they can earn in the market for fish. The availability of extra income from rent seeking in turn allows a fishery of a given size to support more fishers and a larger industry than would be the case absent the availability of rent. And a larger industry means more capacity, adding to the aggregate problem of excess capacity in the global fishing industry.

Regulators, in turn, can increase their bureaucratic profile as more resources are directed toward the populations they regulate. Bureaucracies

21. Charles S. Pearson, *Economics and the Global Environment* (Cambridge: Cambridge University Press, 2000), 440.

have a vested interest in bureaucratic expansion.[22] As the industry pushes for expansion as a means to seek rent more effectively, regulators may well support the push in order to increase their own resources and position within the state, because a bigger industry needs bigger regulatory capacity and thus more regulators.

Industry capture thus can contribute to a vicious cycle in which the industry rent-seeks, leading to excess capacity, and then regulators enable further growth, leading to even greater excess capacity. This further growth in turn may well increase the industry's ability to rent-seek effectively because it is now larger and more people would be negatively affected by a decrease in capacity or income. The development of such a vicious cycle from industry capture of regulators is not particular to fisheries issues. What *is* particular to fisheries issues is the relationship between excess capacity and the resource base. In most industries, the subsidization of excess capacity is inefficient and creates a drain on the national budget but does not directly undermine the industry's future. But because fisheries are a CPR, renewable if managed well but easily depleted if not, industry capture that leads to the subsidization of excess capacity in the aggregate ironically poses a direct threat to the industry's long-term prospects. Regulatory capture is thus problematic for the long term health of fisheries but is nevertheless likely to happen because of the advantages it creates in the present for both the fishers and those who regulate their behavior.

Regulatory Capture and National Fisheries Regulation

The effects of regulatory capture on both domestic and international regulation can be illustrated with a series of brief examples showing both some of the processes through which industry capture of regulators can lead to subsidization and excess capacity and circumstances in which this capture can be attenuated. A systematic review of patterns of subsidy in the global fishing industry is presented in chapter 6, so the illustrations here focus on the link between demands on regulators and decisions to subsidize rather than on subsidies per se.

---

22. Jack H. Knott and Gary J. Miller, *Reforming Bureaucracy: The Politics of Institutional Choice* (Englewood Cliffs, NJ: Prentice Hall, 1987); Carol H. Weiss and Allen H. Barton, eds., *Making Bureaucracies Work* (Beverly Hills, CA: Sage, 1979).

The first example is from the Japanese tuna industry. Japan has a history of strong bureaucratic involvement in industry, in which bureaucracies often have both a significant say in industrial decision making and a vested interest in promoting the industries they regulate.[23] This interest stems both from standard bureaucratic incentives and from a common custom in Japan wherein bureaucrats are hired after retirement by the industries they once regulated.[24] Japan is thus an easy case of industry capture of the fishing bureaucracy.[25] Industry capture in this case is reinforced by a strong food-security discourse, stemming in part from famines after World War II , and by a cultural heritage discourse in a country with a strong fishing tradition.[26]

The combination of industry capture and clear evidence of overcapacity in the 1970s and 1980s led to patterns of subsidization through vessel buybacks and fleet modernization that yielded neither capacity reduction nor an industry capable of supporting itself financially. The government funded buybacks of fishing vessels, in particular older vessels and those owned by the financially weakest fishing companies.[27] But at the same time government assisted in the funding of both modernization of the remaining vessels and construction of new ones. So although in net terms the number of vessels decreased slightly over the two-decade period, fleet modernization in all likelihood led to a significant increase in overall fishing capacity.[28] In fact, the fleet reductions themselves can be seen as an effort to restructure to target higher-value stocks rather than as an

---

23. Susan Carpenter, *Special Corporations and the Bureaucracy: Why Japan Can't Reform* (Basingstoke, UK: Palgrave Macmillan, 2003).

24. Kate Barclay and Sun-Hui Koh, "Neo-liberal Reforms in Japan's Tuna Fisheries? A History of Government–Business Relations in a Food-Producing Sector," *Japan Forum* 20 (2) (2008): 139–170.

25. See, as an example, Polachek, "Politics and Independent Scientific Advice in RFMO Processes."

26. Barclay and Koh, "Neo-liberal Reforms in Japan's Tuna Fisheries?" 144.

27. Marcus Haward and Anthony Bergin, "The Political Economy of Japanese Distant Water Tuna Fisheries," *Marine Policy* 25 (2001): 91–101; Barclay and Koh, "Neo-liberal Reforms in Japan's Tuna Fisheries?" 146.

28. And many of the vessels bought back likely made their way into the Taiwanese and Chinese fishing fleets rather than being taken out of commission, thereby exacerbating rather than ameliorating the overcapacity problem.

attempt to reduce capacity.[29] Despite the combination of modernization and fleet-reduction efforts, however, the industry as a whole operated at a loss throughout the 1980s, when the effects of these programs should have been at their most pronounced.[30]

The Japanese government responded to various industry imperatives in ways that undermined the fishery's financial and ecological sustainability, with a result that institutionalized the requirement of long-term and substantial financial support to the industry. The buybacks intended to decrease overcapacity were undermined by the subsidization for capacity expansion, and the efficiency gains of modernization were undermined by the same excess capacity. The alternatives in this circumstance would either be to allow the industry to shrink to a sustainable size or to use public money to make good the industry's structural losses. The industry itself, as it does in most cases, prefers the latter approach and is able to attract subsidies both through extensive institutional links between industry and government and through the ability to make successful claims on public opinion as guarantors of food security and national culture. The combination of industry capture and public sympathy means that the institutionalization of fishery subsidies in Japan goes largely unchallenged domestically.

A second example of the national policy effects of regulatory capture concerns the policy responses to the creation of EEZs relating to Canada's and Iceland's cod fisheries. Both Iceland and the Canadian maritime provinces have well-developed fishing cultures and economies that have traditionally relied on fishing as a key source of both sustenance and income. The creation of EEZs in both cases brought extensive and lucrative fishing grounds, previously part of the high seas, under national regulation. But the two national fisheries regulators and national governments more broadly reacted to their new economic zones in different ways.

---

29. Kagoshima Prefecture Skipjack and Tuna Fisheries Cooperative Association, *Kagoshima Ken Katsuo Maguro Gyogyô Kyôdô Kumiai Sôritsu Gojûnenshu Shi* (Kagoshima Prefecture Skipjack and Tuna Fisheries Cooperative Association Fifty-Year History) (Tokyo: Suisan Shinshio Sha, 2000).

30. Harry F. Campbell and R. B. Nicholl, "The Economics of the Japanese Tuna Fleet, 1979–80 to 1988–89," in *The Economics of Papua New Guinea's Tuna Fisheries*, ed. Harry F. Campbell and Anthony D. Owen, 39–52 (Canberra: Australian Council for International Agricultural Research, 1994).

The Canadian government reacted by using its new authority over the Grand Banks cod fishery as a regional economic development opportunity. It created incentives both to expand the local fishing industry in the maritime provinces and to encourage the development of a more locally focused industry that would keep the benefits of development in traditional small fishing communities rather than having them become more concentrated in big boats, big companies, and big towns. This approach had the effect of reinforcing the constituency for continued government support of the existing culture of small, single-industry fishing towns. As scientific evidence mounted of strain on the cod stock, the government responded by finding new ways to subsidize the industry in situ rather than restructure it. The cod stock ultimately collapsed to the point where the government was forced to implement a moratorium on cod fishing (which is still in place) but nonetheless still found ways to support existing fishing communities.[31]

The government of Iceland, in contrast, reacted to its new authority by focusing on the ecological sustainability of the fishery rather than on the social sustainability of fishing communities. The ITQ system it eventually developed has had the effect of concentrating fishing capacity in bigger companies with bigger boats at the expense of traditional fishing communities and smaller-scale fishing practices.[32] As a result, the Icelandic cod fishery (along with most of the fisheries within Iceland's EEZ) seems not to be suffering from severe stock decline, and there are no signs of the sort of collapse suffered by Grand Banks cod. This approach and the Icelandic experience with it are discussed further in chapter 9.

What explains the difference between Canada and Iceland despite a similar context? One answer might be the relative economic importance

---

31. For a general overview of this story in the context of the cod fishery more generally, see Mark Kurlansky, *Cod: A Biography of the Fish That Changed the World* (New York: Walker, 1997). See also Lawrence C. Hamilton and Melissa J. Butler, "Outport Adaptations: Social Indicators through Newfoundland's Cod Crisis," *Human Ecology Review* 8 (2) (2001): 1–11, and J. S. Ferris and C. G. Plourde, "Labour Mobility, Season Unemployment Insurance, and the Newfoundland Inshore Fishery," *Canadian Journal of Economics* 15 (3) (1982): 426–441.

32. John D. Wingard, "Community Transferable Quotas: Internalizing Externalities and Minimizing Social Impacts of Fisheries Management," *Human Organization* 59 (2000): 48–57; National Research Council, *Sharing the Fish: Toward a National Policy on Individual Fishing Quotas* (Washington, DC: National Academies Press, 1999), 63, 76.

of the fishery to the national economy. The Canadian fishing industry as a whole is a little less than twice the size of Iceland's in terms of value of fish caught.[33] The Canadian economy, however, is roughly eighty times bigger than Iceland's, so fishing is actually much more central to the Icelandic economy than it is to Canada's. Canada could therefore afford economically to subsidize a financially and ecologically unsustainable fisheries policy as part of a broader pattern of economic support for the relatively poor Maritime Provinces. Iceland could not—its fishery was (and is) a sufficiently central part of its economy that it was forced to think of its fisheries as an economic resource to be nurtured rather than as a cultural artifact to be protected. The result was a focus on sustaining fish stocks rather than on sustaining fishers.

In both the Japanese case and the Canadian case, national governments regulated in the interest of the existing fishing industry rather than in the interest of the country as a whole or of ecological sustainability. If fish stocks could not support existing capacity, the industry's interest was maximized by bringing in resources from national budgets to maintain capacity rather than by taking the ecologically responsible route of reducing capacity to match what stocks could sustain. In Iceland, where the industry is big enough relative to the country that the national government could not afford to let such a pattern of subsidizing overcapacity become institutionalized, industry capture could not draw in an equivalent level of resources from the national budget.

The question for national regulators then becomes what to do with excess capacity. One answer is simply to pay it to sit idle, which is indeed a common practice, as is detailed in chapter 6. Another answer is to try to externalize the problem by finding new fishing grounds and further subsidizing the technology to access fish stocks, which exacerbates the balloon problem we have identified. Although none of these processes makes sense in the context of long-term sustainability of the global fish stock (or even the industry's long-term health), they make economic and political sense in the short-term logic of regulatory capture.

---

33. As of 2002, for example, the total value of industry landings in Canadian ports amounted to roughly US$1.3 billion, whereas the equivalent figures from Iceland were US$829 million. From OECD, *Review of Fisheries in OECD Countries: Country Statistics 2000–2002* (Paris: OECD, 2004).

Conclusions

It might be expected that those who oversee the regulatory processes pertaining to high-seas fisheries would be concerned with protecting the resource in the long term because even if their focus is on the fishing industry's health, that industry will suffer if the fish stocks are not protected. And sometimes they are: the US Fisheries Service, for example, has dramatically improved its management record over the past two decades, in significant part in response to legal pressure from environmental nongovernmental organizations.[34] But structural difficulties often interfere. On the national level, in areas where the fishing industry is a relatively minor part of the national economy, it tends to have disproportionate influence on governmental decision making and is overseen by regulators concerned more with the industry's short-term health than with the long-term protection of the resource.

The international level complicates the issue further, as national representatives make collective decisions within RFMOs. The separation between scientific advice and the political process leads frequently to a situation in which catch levels are set higher than scientists estimate are sustainable and no individual state wants to be left out of access to a collective resource.

Industry capture happens ultimately because governments face multiple pressures with respect to fisheries resources. They face an often unified electoral bloc that would be hurt in the short-term by decreased fishing and are dealing with a CPR that, except within national waters, they cannot restrict access to. They see the advantage of economic development that can be created by accessing a "free" resource that is worth more in the present than in the future to those who seek it. Regulatory capture means that difficult governmental or intergovernmental decisions that might protect global fish stocks are less likely to be undertaken.

---

34. Judith Layzer, "Fish Stories: Science, Advocacy, and Policy Change in New England Fishery Management," *Policy Studies Journal* 3 (1) (2006): 59–80.

5

The Cultures of Fishing Regulation

Regulatory capture of fisheries management by the industry is clearly a stumbling block to the creation of more effective international fisheries regulation. But why does it happen? How is it that the industry is so often able to dominate its regulators at the expense of sound management and society's interest in a healthy marine environment? Regulatory capture is in fact part of a broader tendency in fisheries policy to prioritize the short term over the long term, microregulation over macroregulation, and the interests of fishers over the interests of society at large in a healthy marine environment. There are two sets of reasons for this prioritization, one structural and the other cultural.

Regulatory capture is a common structural problem in many industries because the industry often has a more concentrated interest in its regulation than do other groups. In the case of fisheries regulation, this collective-action problem is exacerbated by the CPR structure of fisheries. But there is another set of reasons as well, involving how we think about the fishing industry and the language with which we talk about fisheries management. Fishing is often more culturally embedded than other industries. And the terms of discourse and analysis of fisheries issues, both in the natural and the social sciences, tends to favor the local over the global and existing institutional structures over the creation of new institutional approaches. This chapter reviews the structural reasons for a bias toward microlevel regulation and fishers' interests and then develops an argument about the way the culture and terms of discourse in fisheries management affects the regulatory approach.

Structures and Institutions

The key structural cause of the frequent bias in favor of fishers' interests is what Mancur Olson called the "logic of collective action."[1] Small groups made up of people with large individual stakes in an issue will generally organize more effectively and therefore be more politically potent than large groups made up of people with smaller individual stakes, even if the overall potential utility to be gained by the larger group is greater. This logic helps to explain the bias within fishery regulators, within governments, and in international cooperative efforts to regulate fisheries.

Within fishery regulators, co-optation by the industry is overdetermined. There are the bureaucratic incentives discussed in chapter 4, in which a bigger industry can generate a bigger bureaucracy. But it is also the case that most people in most countries know and care little about the operation of national (and subnational) fisheries regulators. As a general rule, the only constituencies for which it makes sense to put effort into active lobbying of regulators are the industry itself and environmental groups. The former is often more focused than the latter because environmental groups have many issues other than fish to focus on.[2] The fishing industry is thus the most concentrated and motivated lobby group approaching its own regulator.

Within governments, an industry's concentrated interests in subsidies can offset a more diffuse interest on the part of the broader population not to subsidize. For example, Japan was the largest total fishing industry subsidizer in the mid-2000s. The Japanese government directly or indirectly gave the industry about US$2 billion a year. That works out to just less than $10,000 per person working in the industry, but only about $15 per average Japanese citizen.[3] One can expect that the average fisher

---

1. Mancur Olson, *The Logic of Collective Action: Public Goods and the Theory of Groups* (Cambridge, MA: Harvard University Press, 1965).

2. There are, however, exceptions, particularly when fish species are not just depleted, but also endangered. See, for example, Judith A. Layzer, "Fish Stories: Science, Advocacy, and Policy Change in New England Fishery Management," *Policy Studies Journal* 3 (1) (2006): 59–80, and Thomas Spenser, "Environmental Groups Plan Lawsuit on Petition to Protect Species," *Birmingham News*, 22 April 2011, http://www.biologicaldiversity.org/news/center/articles/2011/birmingham -news-04-22-2011.html.

3. OECD, *Review of Fisheries in OECD Countries 2009: Policies and Summary Statistics* (Paris: OECD, 2010), 47–48.

is going to be willing to put much more effort into getting that $10,000 than the average taxpayer is to save $15. This relationship also helps to explain why small countries with relatively large fishing sectors are often less willing to subsidize excess capacity, as noted in chapter 4. As the cost to the average taxpayer increases, subsidies to the industry are likely to become a more salient political issue to the electorate at large. A person who is unlikely to put much effort into reducing subsidies that effectively cost $15 (a cost that is never directly seen) is much more likely to put in that effort as the cost rises to the hundreds or even thousands of dollars.[4]

Finally, Olsonian logic helps to explain why international fisheries management is dominated by domestic fisheries managers. This dominance might at first glance not need much explanation—national regulators are the logical representatives to RFMOs, given that state representatives are ultimately responsible for policy at these organizations. This logic is reinforced by the likelihood that other parts of government bureaucracies (such as foreign, trade, and environmental ministries) that might have some interest in international fisheries policy are likely to care less than fisheries bureaucracies and are therefore likely to defer to them in decisions about representation at RFMOs. This deference may not happen when fisheries-relevant negotiations happen under the auspices of other international organizations. For example, discussions about fisheries subsidies as part of World Trade Organization (WTO) negotiations have been conducted largely by trade negotiators rather than by fisheries regulators.[5] But for single-purpose international fisheries regulators such as RFMOs, the predominance of national representation by national fisheries regulators is overdetermined.

Another cause of a bias in favor of fishing is to be found in the structure of the economics of the resource, as discussed in several of the preceding chapters. That fish in the wild are generally a CPR means that they are free. There is thus no immediate economic gain in leaving them where they are, and there is economic gain to be had in catching them, which

---

4. Gary S. Becker, "A Theory of Competition among Pressure Groups for Political Influence," *Quarterly Journal of Economics* 43 (1) (2011): 45–65.

5. See, for example, Liam Campling and Elizabeth Havice, "Mainstreaming Environment and Development at the WTO? The Peculiar Case of Fisheries Subsidies," paper presented at the annual meeting of the International Studies Association, New Orleans, 17 February 2010; and WTO, "Negotiations on Fisheries Subsidies," 2001, http://www.wto.org/english/tratop_e/rulesneg_e/fish_e/fish_e.htm.

converts them from an unpriced to a priced resource. Fishers are thus encouraged to overfish. Because the benefit of a healthy fish stock to society is not conveyed to fishers by a price for removing fish from the wild, the absence of this price creates a negative social externality. In other words, whereas the fisher can privatize the profit from removing a fish from the wild, society at large bears a substantial proportion of the cost of the fisher's doing so. This mismatch between private gain and public cost creates a market imperfection that contributes directly to unsustainable fishing.[6]

Governments are also encouraged by this incentive structure to promote overfishing. From their perspective, the fact that fish in the wild are not priced matters for reasons of national accounting rather than for reasons of profit. Contemporary rules for national accounting—for instance, for creating statistics such as GDP—measure economic activity rather than the creation of wealth or well-being.[7] These rules account neither for the effects of economic activity on resource stocks nor for what is created by the economic activity. They give governments an incentive to generate economic activity by depleting resources without providing any mechanism for accounting for the long-term value to society of retaining some of the resource base in an unexploited state. They also do not penalize governments for giving away free resources that governments are supposed to be managing for their citizenry.

This observation is not original—there is a substantial literature within environmental economics on this phenomenon and the problems it creates for sustainable resource management, much of it calling for a new kind of "green accounting" that values resources left for the future.[8] But this phenomenon is nonetheless an important structural consideration driving the overexploitation of global fisheries resources and in particular driving the use of fishing as a form of economic development. In essence, ocean fishing is a way of getting economic development on the cheap

6. Richard Cornes, Charles F. Mason, and Todd Sandler, "The Commons and the Optimal Number of Firms," *Quarterly Journal of Economics* 101 (3) (1986): 641–646.

7. Joseph E. Stiglitz, Amartya Sen, and Jean-Paul Fitoussi, *Mis-measuring Our Lives: Why GDP Doesn't Add Up* (New York: New Press, 2010), 11.

8. See, for example, United Nations Statistical Division, *Integrated Environmental and Economic Accounting: Handbook of National Accounting* (Geneva: United Nations, 1993); Salah El Serafy, "Green Accounting and Economic Policy," *Ecological Economics* 21 (3) (1997): 217–229.

because governments can internalize some of the benefits of fisheries expansion in the form of political benefit from expanded GDP and employment expansion and fiscal benefit from increased taxation, but they also externalize some of the costs in the form of a depleted resource base to society at large.

It must be stressed at this point that although the logic of CPRs is a structural part of the global problem of overfishing, it is neither natural (in the sense that it inheres in the nature of the resource) nor necessary. A resource is common if it has no individual owner and if a group of people has undifferentiated access to it. But ownership is a legal, not a natural, category. In other words, CPRs are unowned because existing legal structures do not create ownership rights for them.[9] Those structures can be changed, as happened when the village commons that gave us the terminology of the commons were gradually enclosed and became private property.[10] It is more difficult to assign property rights to fish than to a pasture, and it is more difficult to account usefully for the value of resources not exploited than for resources in the process of being exploited. More difficult, but not impossible.[11] As discussed in chapter 3, assigning property rights to—in other words, privatizing—fishery resources does not by itself lead to sustainable fishing practices, just as the enclosure of village commons often did not lead to better environmental stewardship. But a lack of attention to the structural problem of fisheries as commons, at both the international level and the domestic level, will undermine effective fisheries management.

The logics of collective action and CPRs drive important arguments about necessary components of the efforts to save global fisheries. Subsidies are one of the mechanisms most directly responsible (through the logic of collective action) for the balloon problem that undermines international fisheries regulation. Although subsidization of the fishing industry

---

9. Elinor Ostrom, Roy Gardner, and James Walker, *Rules, Games, & Common Pool Resources* (Ann Arbor: University of Michigan Press, 1994), chap. 1.

10. See, for example, Garrett Hardin, "The Tragedy of the Commons," *Science* 162 (1968): 1243–1248, and Susan Buck, "No Tragedy of the Commons," *Environmental Ethics* 7 (Spring 1985): 49–61.

11. One way of doing so through survey research is a process called contingent valuation. See, for example, W. Michael Hanemann, "Valuing the Environment through Contingent Valuation," *Journal of Economic Perspectives* 8 (4) (Fall 1994): 19–43.

is less egregious now than it was at times in the past, both the breadth and the scale of the practice remain impressive. Chapter 6 examines the practice and makes the case that a necessary component of dealing with the balloon problem and therefore with the degradation of global fisheries resources is finding a mechanism to effectively phase out the various forms of subsidy. And although privatizing the ocean commons will not by itself prevent overfishing, it will help to temper the balloon problem. Like dealing with subsidies, privatization is not by any means a sufficient condition, but perhaps a necessary one. Chapter 9 looks at mechanisms that assign to fishers property rights to fish and at the same time limits those rights through mechanisms such as ITQs. Chapter 10 discusses the potential for such mechanisms at the international level.

Both of these approaches can usefully address the political, economic, and legal structural causes of overfishing, but they are not by themselves sufficient to address the global problem of overfishing because they do not directly address the practice's cultural embeddedness. It is clear from an economic perspective that limiting subsidies and improving property rights in fisheries management should help to limit overexploitation. Although there has been some movement toward both of these practices recently, there has not been as much as one might expect given the extent to which they make obvious economic sense. The robustness of current practices, ranging from subsidization to a focus on RFMOs, is supported in part by the logics of collective action and CPRs, but it is also reinforced by cultural practices, which are central to understanding global overfishing.

Fishing Culture

Two kinds of cultural practices are part of the process of reinforcing existing norms of fishing and fisheries management: the cultural appeal of fishing and the academic discourse through which fisheries management tends to be discussed. The cultural appeal of fishing itself in turn makes major change in the institutional structure of fishing difficult through both the cultural embeddedness of fishing in communities that are currently or have historically been dominated by the industry and the popular appeal of the idea of fishing and fishing communities to society more broadly.

The cultural embeddedness of fishing in historic fishing communities is fairly straightforward. A fishing industry is often a community activity, involving not only the fishers themselves and their families, but also workers in the port and in fish processing. A culture of fishing pervades the community and is generational.[12] Attempts to limit the fishing industry's activity thus become threats not only to the community's economic underpinning, but also to its identity. Whether such attempts are made by regulators who restrict fishing practices or by marine scientists who argue for tighter quotas, members of the community often interpret them as threats to a way of life rather than as normatively positive efforts either to conserve the resource or to manage the industry rationally.[13] When stocks have clearly crashed, this logic does not apply—if there are no fish, communities generally turn their focus to subsidization to be able to continue their way of life rather than to increased catch limits. But absent unequivocal evidence of decline, attempts to restrain fishing activity can be and often are taken as threats to community integrity.

The industry and others in the fishing community often respond to these regulatory efforts as threats. The industry may respond through organized groups, such as fishers' unions.[14] Interesting examples of responses from fishing communities more broadly can be found in local newspapers in these communities. The *Gloucester Daily Times* from Gloucester, Massachusetts, one of the oldest major fishing communities in North America, covers local fishing issues in depth. Its coverage is often based on careful reporting, but its biases are clear—attempts from outside the community to restrict fishing are perceived as threats to be combated.[15]

---

12. See, for example, Mark Kurlansky, *The Last Fish Tale: The Fate of the Atlantic and Survival in Gloucester, America's Oldest Port and Most Original Town* (New York: Ballantine Books, 2008).

13. N. A. Marshall, "Can Policy Perception Influence Social Resilience to Policy Change?" *Fisheries Research* 86 (2007): 216–227.

14. "Fishing Redundancies Causes Union to Sue," *IceNews: News from the Nordics,* 6 February 2008, http://www.icenews.is/index.php/2008/02/06/fishing-redundancies-causes-union-to-sue; Mary Williams Walsh, "Canada Slashes Cod Quota in Fishing Crisis," *Los Angeles Times,* 25 February 1992, http://articles.latimes.com/1992-02-25/news/mn-2657_1_cod-population.

15. See, for example, Richard Gaines, "Whole Foods Agrees to Halt 'Red' Fish Sales," *Gloucester Daily Times,* 14 September 2010.

The volume of writing on fishing communities and fishing cultures, however, points not only to the internal strength of those cultures, but also to an attraction that those cultures often hold to society at large. This cultural position can be explained in a number of ways. Fishing is in a way the strongest remaining behavioral link we have to human pre-history—fishers are the last of the hunter-gatherers. Fishing has been an activity undertaken throughout history in a consistent enough way that it has found a place in humanity's collective conscious (even though in many instances it is now a far more industrial undertaking than many nonfishers realize). Fishing is associated with the romance of the sea and the lure of fresh food. It is an activity often practiced by nonprofessionals either to supplement diets or for recreation. Anglers may well fail to realize the vast difference in scale and in environmental and ecological impact between what they do and what industrial fishers do.[16]

Fishing traditions (and particular fish) also have cultural resonances in places where the industry has played an important role in the development of local or national identity. Japan and Iceland see catching and eating fish as integral parts of their national culture.[17] Herring in the Netherlands, cod in New England, and salmon in British Columbia are iconic emblems of place.[18] Fishers are the human link to these emblems and therefore make up more than just another industry to be regulated as necessary.

A discourse that defends the right to fish against attempts to regulate can come from points across the political spectrum. For example, the Heartland Institute, a think-tank whose "mission is to discover, develop, and promote free-market solutions to social and economic problems,"[19] including

16. See, for example, Paul Greenberg, *Four Fish: The Future of the Last Wild Food* (New York: Penguin, 2010).

17. Arne Kalland, *Fishing Villages in Tokugawa, Japan* (Honolulu: University of Hawaii Press, 1995), 1; Luca Zarrilli, "Iceland and the Crisis: Territory, Europe, Identity," *Revista Română de Geografie Politică* 8 (1) (2011): 5–15.

18. Christiaan van Bochove, "The 'Golden Mountain': An Economic Analysis of Holland's Early Modern Herring Fisheries," in *Beyond the Catch: Fisheries of the North Atlantic, the North Sea, and the Baltic, 900–1850*, ed. Louis Sicking and Darlene Abreu-Ferreira, 209–242 (Leiden: Brill, 2008); Mark Kurlansky, *Cod: A Biography of the Fish That Changed the World* (New York: Walker, 1997); William L. Lang, "Beavers, Firs, Salmon, and Falling Water: Pacific Northwest Regionalism and the Environment," *Oregon Historical Quarterly* 104 (2) (Summer 2003): 150–165.

19. Joseph Bast, "Welcome to the Heartland Institute," http://heartland.org/about.

environmental protection, criticized recent attempts to institute an ITQ scheme in a US regional fishery and the use of catch shares, another name for ITQs, as a replacement for a general quota and race-for-the-fish system. They did so despite the fact that ITQs are specifically designed as a market mechanism to overcome the market imperfections of general quota systems. The argument against the use of catch shares as a regulatory approach is that it may favor larger fishing companies over individual proprietors—in other words, it threatens the traditional makeup of the US fishing industry.[20] But this argument is not based on the market; it is based on culture.

Cultural embeddedness makes it difficult to engage in policymaking based on economic or ecological calculation. The maintenance of fishing communities becomes an end in itself, even when this end is incompatible with sustainable fisheries management. This dilemma can be overcome when fishing communities find other ways to support themselves, such as whale watching or sportfishing, which generates vastly more income per animal for local operators than commercial fishing does.[21] But such a transition is not feasible for all or even most fishing communities. Whales cannot always be found when wanted; some fisheries (such as anchovies) are not likely to generate much enthusiasm among sportfishers; and many fishing communities are simply too far away from a large enough number of sufficiently wealthy recreational fishers. In the absence of avenues to rebuild fishing communities' economies in ways that retain a relationship with the sea, sustainable management of the world's fisheries will often necessitate the demise of many fishing communities. This necessity is difficult to communicate given current patterns of discourse on the culture of fishing.

The Discourse of Fisheries Management

It is not just the way that fishing is talked about in popular culture that contributes to the embeddedness of the problem of overfishing. The way

---

20. Krystle Russin, "NOAA Eliminates Competitive Fish Harvesting," *Heartland Institute Environment & Climate News* (October 2010): 4.

21. According to the US Fish and Wildlife Service, total sportfishing expenditures in the United States in 2006 (the year of the most recent survey) were $42 billion, more than an order of magnitude greater than the value of the US commercial fishery. US Department of the Interior, Fish and Wildlife Service, and US Department of Commerce, U.S. Census Bureau, *2006 National Survey of Fishing, Hunting, and Wildlife-Associated Recreation* (Washington, DC: US Fish and Wildlife Service, 2006), 9.

that most academics, even those academics whose work specifically focuses on ameliorating overfishing, talk about the problem and potential solutions can institutionalize and reinforce inadequate management processes and institutions. Predominant patterns of discourse lead to a focus on microscale rather than macroscale regulation and management. Decisions by individual scholars to focus on the microlevel are reasonable and generally reflect an academic tendency to speak to existing debates in the literature. But these decisions collectively have the unintended effect of reinforcing an overall pattern of discourse that eschews discussion of macrolevel solutions.

The concept of discourse is used here not in a postmodern sense that focuses on the discourses of modernity as incompatible with emancipatory thought and action.[22] Rather, it is used in a more constrained and specific way to argue that the predominant language with which an issue is discussed has an effect on the content of the discussion.[23] Predominant patterns of academic discourse have an agenda-setting function, except that the agenda in this case is set by a cumulation of authors unintentionally rather than by a single actor who sets out to guide and constrain discussion. Academic work generally sets out to situate itself within existing debates in the relevant discipline, so these existing debates serve as the guideposts for new work. Patterns of discourse therefore tend to become self-reinforcing over time as new work reproduces the frameworks of existing debates. These frameworks can be changed by external events or by groundbreaking new work, but in the course of "normal science" such frameworks have the effect of setting the bounds of discourse.[24]

Terminological choices also constrain scholarship. For example, the academic literature and most of the popular literature tend to speak of the number of fish in terms of weight. Biologists speak of the biomass of a stock, and RFMO quotas happen in tonnes. In both cases, fish are measured by their relationship to other entities—to the ecosystem in the

---

22. In the context of international relations theory, see, for example, Jim George, *Discourses of Global Politics: A Critical (Re)Introduction to International Relations* (Boulder, CO: Lynne Rienner, 1994).

23. See, for example, Stacie Goddard, "When Right Makes Might: How Prussia Overturned the European Balance of Power," *International Security* 33 (Winter 2008–2009): 110–142.

24. The term *normal science* as used in this context is from Thomas Kuhn, *The Structure of Scientific Revolutions* (Chicago: University of Chicago Press, 1962).

first case and to humans as a natural resource in the second. An exception occurred in the literature on incidental dolphin mortality in tuna fishing, where mortality was generally measured and reported as deaths of individual dolphins.[25] There are reasonable technical reasons for this distinction—it is often easier to measure the number of fish caught by weight than to count heads, but dolphins killed incidentally are generally not weighed. They also are large enough that doing a head count is reasonable. The difference in language can affect the way in which information is understood, however.[26] A measurement by aggregate weight suggests that fish are undistinguishable things, whereas a head count suggests that dolphins matter (or at minimum should be thought of) as individuals. It is easier to argue that fish should be caught at MSY but dolphins should be spared when possible if they are described using different terminology.

A focus on microscale regulation is also embedded both in the natural sciences and the social sciences relevant to fisheries management. In the natural sciences, most scholarship is focused either on specific species or local ecosystems. This focus makes sense—it is at the species and local ecosystem scale that most of the questions that fisheries scientists have about fish biology and marine ecology can be answered. And questions about how much fishing effort—or even more so how much capacity—aggregate global fish stocks are able to sustain cannot really be answered effectively from a natural science perspective. Answering these questions requires addressing social scientific questions about the relationship between capacity and effort on the one hand and the effectiveness of regulation on the other. It also requires making ethical and political decisions about applying concepts such as the precautionary principle that the methodology of the natural sciences is unequipped to address.[27]

The microlevel focus of most fisheries science is reasonable in its own context, but it also helps to create a microlevel focus for fisheries

25. For a study in which the two types of measurement discourse (biomass/quotas versus individuals) are juxtaposed, see Elizabeth R. DeSombre, *Domestic Sources of International Environmental Policy: Industry, Environmentalists, and U.S. Power* (Cambridge, MA: MIT Press, 2000).

26. DeSombre plays with this phenomenon in *Domestic Sources*, giving statistics for declining fish populations in head counts rather than weight, thus stressing that they are animals rather than just resources.

27. S. M. Garcia, "The Precautionary Principle: Its Implications in Capture Fisheries Management," *Ocean and Coastal Management* 22 (2) (1994): 99–125.

regulation as well. In important ways, the natural sciences community dominates the academic social sciences community (as distinct from the policymaking community) in the making of fisheries policy, in particular international fisheries policy. Within the academic community, the relative size and level of funding of each of these two research communities vary dramatically. Whereas social scientists who focus on fisheries issues tend to be relatively scarce, with rarely more than one in any given academic department (and none in most departments), fisheries science often has entire dedicated departments.[28] This makes the natural scientists both more numerous and more concentrated, meaning that in interdisciplinary discussions of fisheries management the natural science worldview and discourse is likely to predominate. The tendency of the natural sciences to be better funded than the social sciences exacerbates this phenomenon.[29]

The relative size or funding level of the natural fisheries sciences community relative to the social science community is not in itself problematic—a vast amount of work needs to be done to improve our understanding of marine life and marine ecosystems, and natural science research is generally more expensive to undertake than social scientific research. The problem happens in the translation of research into policy. The discourse of the natural sciences is not designed to capture the social nature of the policy world,[30] so that to allow the natural sciences to dominate the discourse of fisheries policy is to exclude from that policy a full understanding of the particularly social complexities of policymaking. Nor does the discourse of the natural sciences completely dominate fisheries policy—if it did, new experiments in fisheries management such as ITQs would not happen. But this discourse's focus on the effects of fishing on particular

---

28. For example, the University of Washington has a School of Aquatic and Fisheries Sciences, as do the University of Florida and other universities; many other schools (Texas A&M and South Dakota State, to name just two) have a Department of Wildlife and Fisheries Science, and the University of Maryland is one of many with a Department of Fisheries Science.

29. J. B. Lodahl and G. Gordon, "Differences between Physical and Social Sciences in University Graduate Departments," *Research in Higher Education* 1 (1973): 191–213; J. B. Lodahl and G. Gordon, "Funding the Sciences in University Departments" *Educational Record* 54 (1973): 74–82; Jeffrey Pfeffer, "Barriers to the Advance of Organizational Science: Paradigm Development as a Dependent Variable," *Academy of Management Review* 18 (4) (1993): 599–620.

30. Dale Jamieson, "Scientific Uncertainty and the Political Process," *Annals of the AAPSS* 545 (1996): 35–43.

species and ecosystems may reinforce a tendency in fisheries management to rely too heavily on microscale regulation of the interaction of an exogenously given body of fishers with particular stocks rather than on a macroscale regulation of the factors that led to too many fishers in the first place.

This tendency in turn is reinforced by the structure of most RFMOs. As noted in chapter 2, RFMOs generally have integral science commissions—bodies that are designed to provide advice that is seen as scientific and apolitical.[31] The inclusion of these commissions is a feature that distinguishes RFMOs from most other international organizations and has arguably helped RFMOs to set standards and quotas that are significantly stricter than would otherwise have been the case. In other words, the creation of science advisory bodies that member governments accept as independent helps and may be necessary to any success RFMOs have achieved. But at the same time the institutional centrality of natural scientists to the policymaking process means that the process is likely to be heavily influenced by the focus and discourses of the natural sciences.

The discourse within social science treatments of fisheries management also focuses on the microscale, looking at existing patterns of regulation, more than on macroscale management of the supply of fishers. This tendency can be seen from the international level to the local level. One cause of the focus on the microscale in both science and social science is a sort of conservatism of focus.[32] The literatures tend to start with existing socioeconomic and regulatory patterns as a background condition and focus on how to make those patterns work better rather than asking if the patterns need fundamental change. Suggestions for improvements in regulatory processes therefore generally look to making existing structures work better. This type of scholarship reinforces the centrality of

---

31. David Symes, *The Integration of Fisheries Management and Marine Wildlife Conservation*, Joint Nature Conservation Committee (JNCC) Report no. 287 (Peterborough, UK: JNCC, 1998), http://jncc.defra.gov.uk/pdf/RPT287.pdf. Even Tom Polachek's paper about political interference in the scientific RFMO process starts from the assumption that scientific committees are apolitical; see Tom Polachek, "Politics and Independent Scientific Advice in RFMO Processes: A Case Study of Crossing Boundaries," *Marine Policy* 36 (2012): 132–141.

32. Conservatism of focus can be thought of in terms of Kuhn's "normal science," in which most research most of the time focuses on incremental change in existing patterns of study rather than on major change in the focus of study.

those structures, which means it is more difficult to generate a discourse of fundamental change because such a discourse is based on the premise that the fixes proposed for existing structures are insufficient.

Two examples illustrate this point. The first concerns the literature in the global environmental politics research community focusing on international fisheries management. This literature is surprisingly sparse, but much of it focuses either in whole or in part on the role of RFMOs in management.[33] This scholarship rarely discusses the possibility of fundamental institutional change. It is not that the literature on the whole is apologetic for the widespread shortcomings of individual RFMOs or of the RFMO system as a whole. On the contrary, scholars are often quite critical of existing management practices. But to the extent that the literature goes beyond criticism, it often either proposes ways to improve current practices within the existing institutional structure[34] or looks at ways in which the structure is generating improvement from within.[35] Both approaches fail to address the possibility that the basic structure itself is part of the problem.

With respect to international environmental organizations, the central discourse on improved management given existing institutional structures has traditionally been developed from the perspective of neoliberal institutionalism. This approach argues that regulatory failure in international organizations can be understood as a kind of market failure and can be addressed by improving the transparency of international cooperation. This transparency in turn includes improving information availability to all parties, reducing the transaction costs of international cooperation, and improving the contractual environment of that cooperation.[36] These

---

33. See, for example, Michael W. Lodge, David Anderson, Terje Løbach, Gordon Munro, Keith Sainsbury, and Anna Willock, *Recommended Best Practices for Regional Fisheries Management Organizations* (London: Royal Institute for International Affairs, 2007), http://www.chathamhouse.org/publications/papers/view/108473.

34. Ibid., 117–128.

35. Martin Aranda, Paul de Bruyn, and Hilario Murua, *A Report Review of the Tuna RFMOs: CCSBT, IATTC, IOTC, ICCAT, and WCPFC,* EU FP7 Project no. 212188 TXOTX, 2010. http://www.txotx.net/docums/d22.pdf.

36. Peter M. Haas, Robert O. Keohane, and Marc A. Levy, eds., *Institutions for the Earth: Sources of Effective International Environmental Protection* (Cambridge, MA: MIT Press, 1995).

ideas have been applied to the RMFO system, although often with the recognition that even perfectly functioning RFMOs are insufficient to deal with global patterns of overfishing.[37]

Some of the most recent and in-depth research on international fisheries politics uses the perspective of adaptive management, which is designed to explain existing evolutionary patterns of change in organizations rather than to prescribe particular changes. Adaptive management models change in institutional behavior as a set of adaptations to changing circumstances, including the success and failures of past policies.[38] "Adaptive management" is therefore likely a much more realistic description of the evolution of RFMO management than are theories drawing on economics that model policymakers as rational utility-maximizing actors. But at the same time it focuses the study of international fisheries management even more on existing institutions and institutional structures.

The existing literature on international fisheries management is thus developing in interesting ways, and we know much more than we used to about how to make RFMOs work better and how they have evolved in response to the challenge of continued overfishing. The argument made here is not meant as a criticism of the literature and certainly not of individual contributions to the literature. And it must be acknowledged that new research communities, such as those focusing on ecosystems management and marine protected areas, are breaking from this pattern. The focus on making RFMOs fulfill their design function more effectively, however, still predominates in the literature and limits the analysis to those functions for which RFMOs were designed. These functions do not include either macroscale regulation or the regulation of numbers of fishers or fishing capital employed. To the extent that the predominant focus of the literature remains on the RFMO system rather than on the necessity of alternatives, the primary message sent by the community of scholars studying international fisheries regulation will reinforce rather than ameliorate the balloon problem.

---

37. See, for example, Kristina Gjerde, "Editor's Introduction: Moving from Words to Action," *International Journal of Marine and Coastal Law* 20 (3) (2005): 323–344.

38. D. G. Webster, *Adaptive Governance: The Dynamics of Atlantic Fisheries Management* (Cambridge, MA: MIT Press, 2009).

At the local level, trends in the study of the politics of fisheries are skewed even more heavily toward a focus on the maintenance of the status quo. In particular, a recent focus on stakeholders in the literature reinforces current fishers' role in determining both patterns of fisheries management and the distribution of benefits from fishing.[39] The language of "stakeholders" is fundamentally premised on the idea that current participation in fishing brings with it an implicit right either to fish or to be compensated for not fishing. In other words, it emphasizes the rights of the group that, from an Olsonian collective-action perspective, already has the political advantage and has already benefited from open access over any claims that nonfishers might have to the resources of the ocean commons. The logic of the discourse is premised on the idea that because stakeholders have both greater firsthand knowledge about an issue and a personal stake in it, they are best placed to determine a sustainable form of management. The discourse thus necessarily favors a dynamic of extraction over conservation and of the empowered over the disempowered.[40]

But this logic has failed both historically and politically. Historical examples abound of fishing communities opposing restrictions on their ability to fish even in the face of clear scientific evidence of overexploitation.[41] And evidence is rare of efforts by fishers to promote tighter regulation (of themselves rather than regulation designed to exclude others) prior to serious decline in a fisheries stock that is of greater than local scale.[42] This pattern of denial continues to happen—in the Mediterranean, for example, bluefin tuna fishers continue to claim that the entire scientific community is wrong and that there is no real danger of the

39. See, for example, Knut Mikalsen and Svein Jentoft, "From User Groups to Stakeholders? The Public Interest in Fisheries Management," *Marine Policy* 25 (4) (2001): 281–292.

40. See, for example, Judith A. Layzer, *Natural Experiments: Ecosystem-Based Management* (Cambridge, MA: MIT Press, 2008), 71; Robert Phillips, R. Edward Freeman, and Andrew C. Wicks, "What Stakeholder Theory Is Not," *Business Ethics Quarterly* 13 (4) (2003): 479–502.

41. The Georges Bank cod fishery provides just one of many possible examples. See Beth Daley, "Georges Bank Cod Stock on Decline," *Boston Globe*, 22 April 2003, http://www.eurocbc.org/page911.html.

42. T. Hennessey, "Ludwig's Ratchet and the Collapse of New England Groundfish Stocks," *Coastal Management* 28 (2000): 187–213.

collapse of bluefin stocks.[43] In political terms, to the extent that stakeholding contains an element of rent seeking, stakeholders have an incentive to focus their contributions to governance on maximizing resource extraction from society rather than maximizing the long-term sustainability of resource extraction from nature. And the extent to which the global fishing industry is subsidized suggests that stakeholding in fisheries does in fact involve a significant amount of rent seeking.[44]

To focus on a language of stakeholding in the discussion of fisheries policy is therefore to argue for giving the fishing industry institutionalized rent-seeking power over both environmental policy and the national budget. It is essentially to decide ex ante to privilege the socioeconomic interests of fishers over the interest of society at large in sustainable fisheries management. It may still make sense to compensate fishers who are forced out of the industry by the capacity reduction required for sustainable management, although the amount of compensation is subject to a debate that is beyond the scope of this work. But the adoption of the discourse of stakeholding in the study of fisheries regulation biases scholarship, by implicit assumption rather than by explicit argument, in favor of maintaining capacity through subsidization and weak regulation and against the capacity reductions that are necessary for sustainable management.

Conclusions

The challenge of regulating for the common good rather than for the industry's good is a difficulty faced across the economy. The case of fisheries is a particularly difficult one, however. Both the CPR structure of and the inherent limits on fisheries mean that the collective-action problem faced by those concerned about the future of fisheries is more urgent than it is for other industries. Regulatory capture in this case not only redistributes

---

43. Analia Murias, "There Is No 'Possibility of Bluefin Tuna Collapse,' Says Association," *FIS Worldwide*, 8 October 2010, http://www.fis.com/fis/worldnews/worldnews.asp?monthyear=&day=8&id=38544&l=e&special=&ndb=1%20 target=.

44. Bertrand Le Gallic, "Why Is It Difficult for Governments to Move towards Using Market-Based Instruments in Fisheries?" paper presented at the Fifteenth Annual European Association of Fisheries Economists Conference, Brest, 14–16 May 2003, p. 4, http://www.oecd.org/dataoecd/50/27/15354941.pdf.

resources from the population at large to the industry but also threatens the future existence of those resources. Fishers are also culturally different from workers in other industries. We speak about them differently, in both the popular and the scholarly literature. The discourse on fishing often focuses more on cultural maintenance of the industry than on ecological or economic sustainability. Even those parts of the discourse that specifically look at international fisheries regulation are often focused on improving current patterns of governance rather than on fundamentally rethinking existing governance structures. These structural and cultural patterns combine to create a problematic focus on protecting the identity and practices of existing fishers and regulatory processes at the expense of the long-term health of the fisheries and ultimately all those who depend on them.

6

Subsidies and Fishing Capacity

Regulatory capture in the fishing industry and patterns of discourse about the industry that focus as much on cultural maintenance as on sustainable management lead to subsidization of the industry. Subsidies are pervasive, far more so than would be likely without both the regulatory capture and cultural effects; almost every country in the world with a fishing fleet subsidizes it in one way or another. Many different aspects of the fishing industry can receive subsidies: grants, guaranteed loans, or tax incentives can encourage fleet expansion or modernization; discounts or tax deductions on fuel, bait, insurance, or other operating costs can lower the variable costs of fishing; and government grants to secure access to foreign fishing grounds reduce the cost of such access for those who fish there.

The main effect of subsidies is to increase the number of fishers and fishing industries' capacity significantly beyond what the global supply of fish can support and what the market for fish would otherwise maintain. Subsidies are by definition market distorting, by encouraging people to enter or remain in the fishing industry when it would otherwise be too costly for them to do so. They enhance fishers' revenue and thereby make fishing a more profitable activity for those who engage in it than would otherwise be the case. States provide this market distortion because they want to help their domestic fishing fleets. They want to keep people employed, maintain fishing communities, and help provide food for their populations. But market forces serve a useful function: they encourage people to leave the industry when there are not enough fish for fishing to be a profitable endeavor. The market does not work perfectly for any environmental issue and is even worse in the case of fisheries (see chapter 3) than for most other resources, so, if anything, these market forces vastly underreflect the problems experienced in fisheries. But subsidies dramatically and intentionally distort these signals further. That is true even with what most environmentalists would consider "good" subsidies: those intended to encourage fishers to exit the market or those designed to fund scientific research or monitor compliance. All subsidies to some degree distort the market in a way that ultimately contributes to overfishing.

They do so by a number of pathways.[1] For whatever is being subsidized, subsidies lead to overcapitalization—in terms of fishing, an increase in investment in fishing capacity beyond what can be economically supported by the resource. Some of this overcapitalization comes in the form of increased sophistication of technology used, which makes it possible for a smaller number of fishers to catch more fish. Subsidies also allow inefficient use of resources because investment does not have to be supported by earned revenue. Through all these pathways, subsidies encourage overharvesting by reducing the marginal cost of fishing. Subsidies also encourage overconsumption because the price of fish does not reflect the actual cost of catching them, so people come to eat more fish or find other uses for them, such as in animal feed or in processed foods, than they otherwise would. These effects collectively add up to depletion of a CPR, thereby depriving others (or the ecosystem) of fish that would otherwise remain available.

Fisheries subsidies have existed for centuries. In 1640 in the Massachusetts Bay Colony, those who fished were given military exemptions and tax wavers for fishing vessels and gear.[2] British subsidies in the 1700s sought to develop the Scottish fishing industry and increase British fisheries competition in foreign markets.[3] Developing-country fishery subsidies are more recent but have nevertheless existed for decades. In the early 1970s, for example, Chile and Brazil used exemptions from taxes and import duties as a tool for development of domestic fisheries, and Peru created a program for investment in fisheries infrastructure and equipment.[4]

1. Gareth Porter, *Fisheries and the Environment: Subsidies and Overfishing, towards a Structured Discussion* (Geneva: UNEP, 2001).

2. William E. Schrank, *Introducing Fisheries Subsidies*, FAO Fisheries Technical Paper no. 437 (Rome: FAO, 2003), 14.

3. Anna Gambles, "Free Trade and State Formation: The Political Economy of Fisheries Policy in Britain and the United Kingdom circa 1780–1950," *Journal of British Studies* 39 (3) (2000): 288–316.

4. Deepali Fernandes, *Running into Troubled Waters: The Fish Trade and Some Implications*, Evian Group Policy Brief (Geneva: Evian Group, November 2006), http://papers.ssrn.com/sol3/papers.cfm?abstract_id=1138271.

The major expansion of the global fishing industry beginning in the 1950s was fueled by subsidies. Investment in the creation or expansion of distant-water fishing fleets that could process catches on board and remain at sea allowed states that had already achieved dominance in high-seas fishing to expand their reach and seek new fish stocks. Subsidies allowed new ships to be built or financed at low cost and to be equipped with the latest technology. The expansion of domestic fisheries jurisdiction via the recognition of EEZs led to the creation of many new subsidy programs as states (often those without a major fishing presence) encouraged the development of domestic industries in order to take advantage of the resources they were newly able to control (chapter 7 provides some examples of these programs and the difficulties they created). Fishing subsidies are high: compared to subsidies that support the production of other animal protein sources, fish joins only beef at the top of the list globally.[5]

Because of the variety of ways to define or estimate subsidies to the fisheries sector, it is impossible to give a systematic overview of fisheries subsidies. This chapter, therefore, discusses the issue conceptually, examining the type and range of subsidies (and subsidizers) and their effects, with some extended examples designed to illustrate the difficulties that subsidies create for global (and even local) management of fisheries. They primarily create additional fishing capacity in a situation in which there is already more fishing capacity than global stocks can support, but they also encourage a related political dynamic. They create a perpetual constituency for further support (and for the maintenance of a local or global fishing fleet far beyond what existing stock size can justify), which creates a vicious cycle. The biggest problem with subsidies ultimately is that they are difficult to remove.

What Are Subsidies? What Is Their Magnitude?

There is no one definition of what subsidies are. The WTO's 1994 Agreement on Subsidies and Countervailing Measures defines them as "financial contributions provided by governments," including in that definition

---

5. Matteo Milazzo, *Subsidies in World Fisheries: A Reexamination*, World Bank Technical Paper no. 406 (Washington, DC: World Bank, 1998).

transfers of funds (such as grants or loans) or potential transfers (as in loan guarantees); foregone government revenues (such as tax preferences); governmentally supplied goods and services (but not general infrastructure); and price or income-support programs.[6] Even general infrastructure, excluded from the WTO definition, can have the effect of a subsidy. To the extent that fishers make use of port facilities that license fees do not contribute to creating or that user fees do not fully support, the government is making a contribution to their ability to fish.

The wide variety of types of measures that can be considered subsidies makes difficult any comparisons of levels of subsidization across countries or fisheries (or even across studies). We can nevertheless get a sense of the magnitude of subsidization by looking at some global estimates and what they include or omit. A 2006 University of British Columbia study, which excludes fuel subsidies in its parameters, estimated global fisheries subsidies at US$26 billion each year;[7] the same research center estimated that fuel subsidies for fishing are approximately US$4–8 billion annually,[8] for a total of US$30–34 billion. Other estimates are more modest (but omit some relevant types of subsidies): a widely cited World Bank study from 1998 estimated global annual fisheries subsidies of US$14–20 billion,[9] and the World Wildlife Fund (WWF) estimated subsidies in the late 1990s at a minimum of US$15 billion.[10]

Another way to try to understand the scope and scale of fisheries subsidies is to compare them to the industry's revenue. The WWF estimated

---

6. WTO, Agreement on Subsidies and Countervailing Measures (1994), Article 1.1.

7. Ahmed S. Kahn, Ussif Rashid Sumaila, Reg Watson, Gordon Munro, and Daniel Pauly, "The Nature and Magnitude of Global Non-fuel Fisheries Subsidies," in *Catching More Bait: A Bottom-Up Re-estimation of Global Fisheries Subsidies*, ed. Ussif Rashid Sumaila and Daniel Pauly, 5–37, Fisheries Centre Research Reports, vol. 14, no. 6 (Vancouver: Fisheries Centre, University of British Columbia, 2006).

8. Ussif Rashid Sumaila, Louise The, Reg Watson, Peter Tyedmers, and Daniel Pauly, "Fuel Subsidies to Global Fisheries: Magnitude and Impacts on Resource Sustainability," in *Catching More Bait*, ed. Sumaila and Pauly, 38–48.

9. Milazzo, *Subsidies in World Fisheries*.

10. WWF, *Hard Facts, Hidden Problems: A Review of Current Data on Fisheries Subsidies* (Washington, DC: WWF, 2001).

in 2004 that fishing subsidies totaled nearly 20 percent of revenue in the fishing industry.[11] The World Bank 1998 estimate calculated subsidies at 17–25 percent of all fishing revenues.[12] Other estimates give an even greater role to subsidies. A 1993 FAO study estimated that global fisheries costs were greater than revenues by 78 percent (US$54.5 billion), which could have been accomplished only by subsidies in that amount.[13] Subsidies just to bottom trawl vessels probably account for 25 percent of the landed value of the catch from these vessels.[14]

Who Subsidizes?

Because there are no uniform definitions of subsidies, it can be difficult to make precise conclusions about specific levels of subsidies by states, but the bulk of subsidy use is concentrated among a small number of states. The WWF estimated that 90 percent of all fisheries subsidies in 1996 came from seven states or regional economic integration organizations: Japan, the EU, the United States, Canada, Russia, Korea, and Taiwan.[15] India also has significant fisheries subsidies,[16] and China's subsidies are probably dramatically underreported.[17]

By most measures, Japan subsidizes the greatest amount, providing more than US$2 to $3 billion[18] (and by some estimates more than $5

---

11. David K. Schorr, *Healthy Fisheries, Sustainable Trade: Crafting New Rules on Fishing Subsidies in the World Trade Organization*, WWF Position Paper and Technical Resource (Washington, DC: WWF, June 2004), 10–11.

12. Milazzo, *Subsidies in World Fisheries*.

13. FAO, "Marine Fisheries and the Law of the Sea: A Decade of Change," special chapter (revised) of *State of Food and Agriculture 1992*, Marine Fisheries Circular No. 853 (Rome: FAO, 1993).

14. Ussif Rashid Sumaila, Ahmed Kahn, Louise The, Reg Watson, Peter Tyedmers, and Daniel Pauly, "Subsidies to High Seas Bottom Trawl Fleets and the Sustainability of Deep Sea Benthic Stock," in *Catching More Bait*, ed. Sumaila and Pauly, 49–53.

15. WWF, *Hard Facts, Hidden Problems*.

16. Kahn et al., "The Nature and Magnitude of Global Non-fuel Fisheries Subsidies," 5.

17. WWF, *Hard Facts, Hidden Problems*.

18. Ibid., 17.

billion[19]) annually to its fishing industry. This support goes primarily to infrastructure and capital improvements but has also included more than $500 million in vessel insurance, $200 million in access agreements, and another $270 million to help improve harvesting techniques.[20] Subsidies to the Asian fishing industry collectively may come to more than $16 billion.[21]

The EU states individually provide subsidies that collectively total more than $3 billion. Some states, such as Denmark and Malta, provide subsidies that are greater than the value of the fish caught. Spain and France heavily subsidize fuel. Another highly subsidized category across Europe (especially Spain, Italy, France, and Denmark) is vessel construction and equipment, which has encouraged the increasing size and modernization of the fishing fleet.[22]

The United States subsidizes across all categories (except for fisheries-specific unemployment insurance[23]), providing at least $1.2 billion in subsidies each year.[24] Canada reports approximately US$800 million in fisheries subsidies annually, focusing much of its subsidy on income support and unemployment insurance specific to the fishing industry. It also has some marketing and price-support programs and subsidizes fisheries management and conservation programs.[25]

Developing states are also increasingly turning to subsidies to develop their fishing fleets. Some of the developing countries that provide the highest subsidies currently are in South America: Peru, Chile, Brazil, Argentina, and Uruguay all highly subsidize their fleets. Latin America and the Caribbean together account for annual subsidies of at least $3 billion.[26]

---

19. Oceana, *Global Fisheries Subsidies Regional Breakdown*, as derived from *Catching More Bait: A Bottom-Up Re-Estimation of Global Fisheries Subsidies*, 2nd version, ed. Ussif Rashid Sumaila and Daniel Pauly, Fisheries Centre Research Reports (Vancouver: Fisheries Centre, University of British Columbia, 5 June 2007), 1, http://oceana.org/en/our-work/promote-responsible-fishing/fishing-subsidies/learn-act/more-on-fisheries-subsidies.

20. Milazzo, *Subsidies in World Fisheries*, 18–20, 39.

21. Oceana, *Global Fisheries Subsidies Regional Breakdown*, 1.

22. Ibid., 1.

23. WWF, *Hard Facts, Hidden Problems*, Annex I, A-13.

24. Oceana, *Global Fisheries Subsidies Regional Breakdown*, 4.

25. WWF, *Hard Facts, Hidden Problems*, Annex I, A-3.

26. Oceana, *Global Fisheries Subsidies Regional Breakdown*, 2.

An alternative to considering absolute amounts of fishing subsidization is to put those amounts into context. One option considers "subsidy intensity,"[27] comparing the subsidy amount to the value of the landed catch. By this measure, some states—such as Micronesia, the Ivory Coast, Mauritania, Algeria, Antigua and Barbuda, Brazil, Guatemala, and Grenada—provide subsidies at nearly twice the value of landed catches.[28]

It is no surprise that subsidies have increased the size and capacity of the fishing fleets in countries that use them; such an increase is, after all, the whole point of subsidizing. But the economic inefficiency should also be apparent, especially in places that grant subsidies that are higher—sometimes much higher—than the value of the fish landed by subsidized fishers. It is clear in these contexts that subsidies are not about producing economic value, but about political support to certain communities or economic sectors. In many of the places where the fishing industry is subsidized, these subsidies do not make economic sense.

What Is Subsidized?

There are as many ways to classify subsidies as there are definitions of what constitutes a subsidy. Some researchers consider the extent to which subsidies are explicit (involving specific government outlays of funding) or implicit (in which the government suppresses supply prices).[29] Others refer to whether subsidies are direct or indirect.[30] Gareth Porter classifies them by whether they are border or domestic instruments.[31] Yet others categorize subsidies as "good" (assistance with surveillance and management), "bad" (assistance for such things as vessel construction or equipment upgrades), and even "ugly" (those subsidies that, like boat buyback programs or community development programs, are hard to

---

27. Sumaila and Pauly, eds., *Catching More Bait*.

28. Oceana, *Global Fisheries Subsidies Regional Breakdown*, 3.

29. L. Reijinders, "Subsidies and the Environment," in *Producer Subsidies*, ed. R. Gerritse, 111–121 (London: Pinter, 1990).

30. UNEP, Division of Technology, Industry, and Economics, "Subsidies in Argentine Fisheries," UNEP Fisheries Workshop, Geneva, 12 February 2001, http://www.unep.ch/etb/events/events2001/fishery/fisheryArgentina.pdf.

31. Gareth Porter, *Fisheries Subsidies, Overfishing, and Trade*, Environment and Trade Report no. 16 (Geneva: UNEP, August 1998).

judge without knowing the context).[32] A United Nations Environment Programme (UNEP) study provides the most useful starting point for classifying subsidies;[33] it is the basis for the initial categorization used here. Although all subsidies have the effect of distorting the market and thereby providing incorrect price signals to participants in the industry, some types of subsidies have more pernicious effects than others in terms of increasing and maintaining excess capacity in the fishery. Subsidies are discussed here in rough order of their importance in contributing to problems of increasing capacity and decreasing management opportunities for fisheries generally.

Capital Costs

This category of subsidy is what most people think of first when considering fisheries subsidies. The government encourages increase in or modernization of fishing fleets via grants, loan guarantees or loans at below-market rates, or tax incentives designed to make it easier for vessel owners to build, buy, and equip vessels to fish. Subsidies for capital costs increase the number of vessels fishing and accelerate the acquisition of gear and the adoption of new technology. These subsidies do the most to increase fishing capacity directly.

Shipbuilding subsidies are the most obvious of the capital-cost subsidies. In the early to mid-1990s, Germany, Italy, Japan, Norway, Spain, and the United Kingdom were among those states with the most significant subsidies for building fishing vessels. These six states together accounted for one-third of the new fishing vessel construction during that period.[34]

National subsidy programs rarely discriminate by where the vessels are fishing and may therefore worsen problems of overcapitalization in domestic fisheries that were already at capacity. In international fisheries, national subsidy programs are frequently intended to allow the subsidizing state's fishers to compete better globally. But other states fishing in

---

32. Kahn et al., "The Nature and Magnitude of Global Non-fuel Fisheries Subsidies."

33. UNEP, *Analyzing the Resource Impact of Fisheries Subsidies: A Matrix Approach* (Geneva: UNEP, 2004).

34. J. Fitzpatrick and C. Newton, "Assessment of the World's Fishing Fleet 1991–1997," Greenpeace, 5 May 1998, http://archive.greenpeace.org/oceans/globaloverfishing/assessmentfishingfleet.html.

an area are also likely to subsidize their vessels, which leads to a fishing version of an arms race.

Subsidizing capital costs can have a dramatic impact both on a national fishing fleet and on the fishery in which it operates. For example, in the 1950s and 1960s Canada highly subsidized its high-seas fishing fleet in the Atlantic via grants and loans to increase Canadian ability to compete with European vessels for cod catches. As a result, the Canadian trawler fleet increased eighteenfold. By its own estimates, the fleet reached twice the capacity that was economically sustainable by 1970, and catches began to decline significantly soon thereafter.[35] Senegal similarly wanted to be able to increase its catches relative to others in the waters off its coast (even before EEZs were codified under UNCLOS), so it began to subsidize engines sold to small-scale fishers. Although none of these vessels was motorized in 1960, four thousand of them were by 1974. Subsidies also assisted in the Senegalese fishers' adoption of purse seine nets during the 1970s.[36]

Subsidies to capital costs allow more people to purchase vessels and those that are purchased (or updated) to be larger and more technologically sophisticated than they would be without the subsidies. The increased number, size, and sophistication of fishing vessels enables people to fish farther from home (and to move to follow available fish stocks or to escape existing regulatory restrictions), to catch a greater amount of fish, and to do so more easily.

Operating Costs

Assistance with the variable costs of operating fishing vessels is another reasonably obvious category of subsidy that also directly increases the level of fishing intensity. This category includes subsidized or guaranteed insurance, fuel tax exemptions (or otherwise below-market fuel costs), deferral of income tax, bait subsidies, compensation for damaged gear, preferable rates (or interest deductions) on loans for operating costs,

---

35. Porter, *Fisheries Subsidies, Overfishing, and Trade.*

36. Karim Dahou, Enda Tiers Monde, and M. Dème, with the collaboration of A. Dioum, "Support Policies to Senegalese Fisheries," in *Fisheries Subsidies and Maritime Resources Management: Lessons Learned from Studies in Argentina and Senegal,* ed. UNEP, 25–53 (Geneva: UNEP, 2002).

subsidies for transportation, any other government assistance with fishing operations,[37] and related financial assistance.

Ocean trawlers on the high seas are particularly highly subsidized, at what one study estimates is US$152 million annually, more than half of which is in the form of fuel subsidies. Without these subsidies, "the bulk of the world's bottom trawl fleet operating in the high seas will operate at a loss (unable to fish)."[38]

These types of subsidies make the operations of specific fishing trips less expensive for fishers than they would otherwise be (as differentiated from the capital-cost subsidies, which have the same effect no matter how many days of fishing a vessel undertakes). They therefore increase, among other things, the range at which it makes sense to fish. They may also allow for an increase in capacity (such as a more powerful engine)[39] and for increased use of ice or refrigeration, which can expand effort. Fuel subsidies may increase vessel range and thereby expand fishing vessels' reach.

Fishing Access

Another category of subsidy involves the payment for access to foreign fishing areas. This kind of subsidy is really a subset of operating costs. Because states can control access to their EEZs, they frequently charge others for access; distant-water fishing fleets that want legal access to EEZs can gain it only through the payment of such fees. This kind of payment is most frequently made by governments on behalf of vessels that fly their flag, although it is not always framed entirely as a direct-access payment. For example, the EU's access agreement with Mauritania covering the years 2008–2012 includes €1 million per year to support a national park, out of an average annual payment of just more than €76 million.[40] Access agreements between the EU and Iceland and Norway involve granting reciprocal access to each other's fisheries without any money changing

37. UNEP, *Analyzing the Resource Impact of Fisheries Subsidies*, 10.

38. Sumaila et al., "Subsidies to High Seas Bottom Trawl Fleets," 49.

39. J. R. Beddington and R. B. Rettig, *Approaches to the Regulation of Fishing Effort*, FAO Fisheries Technical Paper no. 243 (Rome: FAO, 1983); James R. McGoodwin, *Crisis in the World's Fisheries: People, Problems, and Politics* (Stanford, CA: Stanford University Press, 1990), 172.

40. European Commission, "Mauritania: Fisheries Partnership Agreement," updated 17 February 2012, http://ec.europa.eu/fisheries/cfp/international/agreements/mauritania/index_en.htm.

hands. Most access agreements, however, do generally spell out specific fees for types of vessels and type of fish caught.

The most prominent of these agreements are between the EU (primarily on behalf of Spanish and Portuguese fleets, although French and Italian fleets also participate) and West African states, and between the United States and Pacific Island states. As of 2009, the EU paid approximately €226 million (approximately US$320 million) in total annual fees for fisheries access.[41] The United States in the 1990s did not negotiate specific access agreements with Pacific Island states but did give a $15 million grant; as a result, US tuna fishing vessels paid only about one-quarter of what vessels from other distant-water fleets were charged to fish in this area.[42]

These agreements can be extremely important sources of revenue for the developing states that sign them. For instance, the Pacific Island states Kiribati and Tuvalu earn nearly 42 percent of their GDP from access fees.[43] The agreements nevertheless are designed primarily to benefit the fishers granted access. The European Commission acknowledges that these agreements "enable the European industry to count on an additional supply of fish [and] provide employment for many European workers."[44]

But access agreements as they currently function have problematic effects even beyond the subsidy's direct effects. These problematic effects come from the lack of governance capacity on the part of the developing states with which they are negotiated and the problematic nature of the agreements themselves. As a result, such access agreements contribute more actively to the depletion of the fisheries in the waters to which they apply than would be the case with the vessels in question fishing in their national waters.

The first sustainability problem created by many access agreements is with developing states' ability and willingness to monitor foreign vessels

41. European Commission, "Bilateral Fisheries Partnership Agreements between the EC and Third Countries," 16 July 2009, http://ec.europa.eu/fisheries/cfp/external_relations/bilateral_agreements_en.htm (accessed on 29 July 2009).

42. R. Gillet, M. A. McCoy, and D. G. Itano, *Status of the United States Western Pacific Tuna Purse Seine Fleet and Factors Affecting Its Future*, Publication no. 02-01 (Manoa: School of Ocean and Earth Science Technology, University of Hawaii, 2002).

43. Fernandes, *Running into Troubled Waters*, 8.

44. European Commission, "Bilateral Fisheries Partnership Agreements Between the EC and Third Countries."

and effectively enforce existing regulations. Foreign fishing vessels take advantage of that lack of oversight by catching species beyond those that they are authorized to catch. Licenses are for a specific type of fish, but there is little monitoring of what is actually caught, so ships can misrepresent what they intend to catch. For instance, only 8 percent of the catch by Italian-flagged vessels with shrimp licenses in the waters of Guinea-Bissau in 1996 was shrimp; the majority was demersal fish and cephalopods. Portuguese-registered shrimp vessels similarly had catches that were only 27 percent shrimp.[45] Because there are few limits on bycatch and those that exist are not monitored or enforced, fishing vessels can obtain the least expensive license and then fish for whatever they want.

This lack of monitoring capacity also influences the nature of the agreements themselves. States that are charging for access are unlikely to have sufficiently researched the status of the stocks in question and are therefore unlikely to be able to determine what a sustainable vessel capacity would be. These agreements also frequently include no restrictions on the amount of fish that can be caught; access and license fees are based on number of vessels and vessel tonnage, but not on how much the ships actually catch, allowing "the EU fleets to harvest an essentially unlimited volume of resources for one pre-fixed license payment."[46] US agreements with Pacific Island states often have a similar lack of quota.[47] The fishing vessels or foreign states also routinely do not report any information about catches to the countries in whose waters they are fishing[48] and do not allow observers from the coastal countries on board,[49] which perpetuates the lack of information these countries have about the health of their own fisheries.

Because the revenue that these developing states receive in return for access is the reason they are willing to negotiate these deals, they may grant greater access to foreign vessels than they would otherwise choose

---

45. Vladimir M. Kaczynski and David L. Fluharty, "European Policies in West Africa: Who Benefits from Fisheries Agreements?" *Marine Policy* 26 (2002): 75–93.

46. Ibid., 78.

47. UNEP, *Analyzing the Resource Impact of Fisheries Subsidies*, 24.

48. Kaczynski and Fluharty, "European Policies in West Africa," 78.

49. Ibid., 87.

to allow if they were managing for the health of the resource. In addition, they may be reluctant to decrease access (and therefore reduce the funding available) even as resources become depleted.[50] Although agreements are entered into voluntarily, the short time horizons of the government officials seeking foreign exchange may overcome their interests in long-term protection or development of their coastal fisheries.

Access agreements are thus subsidies from governments of developed countries to their own fishers that provide the fishers free access to fishery resources they would otherwise have to pay for access to. This subsidy encourages fishing beyond what would be economically sustainable by those engaging in it. And the financial need of the developing countries that grant this access in return for payment leads them to underregulate the terms of access, while the inadequate administrative capacity that stems from poverty prevents enforcement of even those lax terms.

Income Supports

Another way governments subsidize fishers is to ensure that fishers receive a reliable wage, regardless of fishing conditions or regulations. These income subsidies include compensation for closed seasons or unemployment insurance (especially when unemployment insurance or resources to those in the fishery go beyond what might be offered in other industries).

The most common form of income support is a government grant to compensate for closed seasons or areas (also called "laying up" insurance); these subsidies often accompany a temporary fishing moratorium. European states have been the predominant users of this type of fishing income support. Such programs provide needed income to those who find themselves suddenly unable to fish, and they are thus considered to be humane ways to address government policies to protect fisheries. But practices of this sort discourage capacity reductions when fishing conditions might otherwise make such reductions desirable. Because compensation for closed seasons enables fishers to remain in the industry, it contributes to the pressure to allow increased fishing access, if not now, then

---

50. Marine Resources Assessment Group of the UK Department for International Development, *Fisheries and Access Agreements*, Policy Brief no. 6 (London: MRAG, n.d.).

sometime in the future. An EU report on these programs suggested that they generally fail to change fishing behavior in the long term.[51]

In open-access fisheries, income supports are likely to increase fishing effort.[52] People may choose to remain in the fishing industry rather than seek other employment, and, worse, others may choose to enter because of the generous compensation when fishing is not allowed or available. Empirical studies bear out this conclusion. Unemployment insurance that pertained specifically to fishers in the Atlantic provinces in Canada was found to contribute to an increase in the number of fishers in the region.[53] Models suggest that the effect depends on the type of payment. For instance, unemployment insurance payments that are a function of income have the effect of increasing fishing effort.

Although additional unemployment insurance beyond standard national programs can contribute to this problem, "laying up" subsidies, which are framed as temporary measures until the fishery can sustain normal fishing activity, are especially harmful in this respect. The biggest problem with income-support subsidies is that they allow or encourage fishers to remain in the industry. They not only prevent an economically rational diminishment of the size of the fishing community but lead to increasing political pressure on managers to reopen a closed fishery or increase allowable catches.[54]

Price Supports

Governments may also intervene in the market to create a minimum price for fish products; doing so ensures that no matter how much fish is landed,

---

51. European Court of Auditors, "Common Policy on Fisheries and the Sea," *Official Journal of the European Communities* (15 December 1992): 107–115; UNEP, *Incorporating Resource Impacts into Fisheries Subsidies Disciplines: Issues and Options, a Discussion Paper*, UNEP/ETB/2004/10 (Nairobi: UNEP, 2004), 23, http://www.unep.ch/etu/Fisheries%20Meeting/IncorResImpFishSubs .pdf.

52. Erik Poole, "Income Subsidies and Incentives to Overfish," in *Microbehavior and Macroresults: Proceedings of the Tenth Biennial Conference of the International Institute of Fisheries Economics and Trade Presentations*, 1–13 (Corvalis, OR: Institute of Fisheries Economics and Trade, 2000).

53. J. S. Ferris and C. G. Plourde, "Labour Mobility, Season Unemployment Insurance, and the Newfoundland Inshore Fishery," *Canadian Journal of Economics* 15 (3) (1982): 426–441.

54. UNEP, *Analyzing the Resource Impact of Fisheries Subsidies*, 40.

the price will remain high. Governments may use price supports to keep domestic prices above the global price. This kind of support involves market manipulation—withdrawing or holding back fish from the market to increase the prices of those being sold. In other words, the government designates a price at which it will buy fish and take it off the market if the price falls too low. Fishers thus know they are guaranteed a price floor and do not respond to the signal that fish prices are decreasing by reducing their fishing effort. The EU is the primary practitioner of this form of subsidy.[55]

The Norwegian experience with price supports illustrates some of the difficulties they can cause. Norway subsidized its fishing industry in the 1970s and 1980s primarily through price supports that increased from 5.5 to 16 percent of the price of cod between 1978 and 1985.[56] The price supports contributed, scholars argue, to a decline in the relevant stocks during that period because they encouraged fishers to fish even when the price on the global market was low.[57]

Because prices are a signal (even if an imperfect one) that can help fishers determine fishing effort, ensuring that prices do not drop even if supply exceeds demand encourages fishers to fish beyond the level they would otherwise consider to be profitable. Price supports historically have increased fishing effort, leading to overexploitation.[58] Even in fisheries with catch regulations, the incentives created by price supports entice fishers to put pressure on managers to increase catch limits (or to decide to illegally circumvent the limits)[59] because the fishers know that they are guaranteed a certain price for their catches.

Fishing Infrastructure

In order to be able to catch and land fish safely, especially on an industrial scale, port facilities are required. These facilities are frequently provided

---

55. OECD, *Transition to Responsible Fisheries, Government Financial Transfers, and Resource Sustainability: Case Studies*, AGR/FI(2000)/10/FINAL (Paris: OECD, 2000), 98.

56. OECD, *Transition to Responsible Fisheries: Economic and Policy Implications* (Paris: OECD, 2000), 146–147.

57. P. Salz, *The European Atlantic Fisheries* (The Hague: Agricultural Economics Research Institute, 1991); O. Flaaten and P. Wallis, *Government Financial Transfers to Fishing Industries in OECD Countries* (Paris: OECD, 2000).

58. UNEP, *Analyzing the Resource Impact of Fisheries Subsidies*, xii.

59. Ibid., 42.

through public (government) funding. Moreover, these ports are often improved over time, especially in ways that make them more useful for industrial fishing—for instance, through dredging to enable servicing of larger vessels and through the creation of harbor facilities and areas for mooring.

Most ports are built and modernized by governments, so to this extent governments around the world regularly subsidize this aspect of fishing behavior. If ports and facilities are not funded by user fees, these government actions decrease the costs of fishing. Because these facilities have in most cases been provided by governments from the beginning, this subsidy is hidden; it has consistently lowered the cost to fishing fleets over time to such an extent that it is rarely even considered to be a subsidy. Government provision of other infrastructure such as lighthouses, weather reports, satellite navigation, and so on without recoup of the costs through user fees also has the effect of subsidizing the fishing industry.

Japan has had a particularly ambitious program of port expansion for the fisheries sector. Between 1994 and 1999, it allocated US$5 billion annually to port expansion and maintenance.[60] Port construction and access without user fees can also make an area's waters more appealing to distant-water fishing nations and thereby contribute to increased exploitation of new fishing areas. Japan's funding of port construction in Indonesia[61] may have been driven in part by the perceived need for improved access for Japanese fishing fleets.

Even more interesting is foreign investment in fishing infrastructure in the developing countries in whose waters the investors fish or hope to gain fishing access. For example, Spain has invested in ports in Ecuador,[62] and Japan has in Ghana.[63] This type of infrastructure subsidy operates simultaneously as an access subsidy as well if it comes with preferential access for the donor country's fishers or if it simply makes these ports more usable for donor nationals who are otherwise fishing in the region.

---

60. Ibid., 19.

61. Ibid.

62. See, for example, "Government to Promote Manta Port among Spanish Investors," *The Construction Gateway*, 28 December 2009, http://www.construpages .com/nl/noticia_nl.php?id_noticia=1191&language=en.

63. Japanese Ministry of Foreign Affairs, "Summary of the Joint Mission for Promoting Trade and Investment for Africa (the Central and West Mission)," 8 October, 2008, http://www.mofa.go.jp/announce/announce/2008/10/1185334_1060 .html.

And, of course, it makes further development of the local fishing industry possible and thereby contributes to increased global capacity.

Fishing infrastructure subsidization is a mostly hidden type of subsidy that affects anyone that uses the infrastructure. Governments can and do focus on improving these services to make their fishers more competitive on a global market. In some cases, this government intervention might make industrial-scale fishing possible where it otherwise would not have been, but this infrastructure is rarely used by fishers only. It should be considered a form of subsidy, but it is difficult to determine what to attribute specifically to fishing, and it has a much smaller effect than the preceding forms of fishing subsidy.

Management Services

Governments also contribute fishery-related services that provide beneficial information to those fishing in the area: monitoring stocks and providing surveillance on fishing behavior; undertaking research; making and enforcing domestic rules about fishing, and participating in RFMOs. Government-funded feasibility studies on fisheries resources[64] similarly provide valuable information to the fishing industry that the industry does not have to pay for, thus increasing catches for those who do not bear the cost of producing that information. The FAO estimates that subsidies for this purpose account for more than one-third of overall Organization for Economic Cooperation and Development (OECD) fisheries subsidies.[65]

This category of subsidy is often ignored by antisubsidy efforts (such as those within the WTO) because many of the latter management efforts are in the service of increasing sustainability of harvesting. Some consider this provision of services to be an "implicit" subsidy.[66] But because the fishers in these states do not have to bear the costs of these activities themselves, these services are effectively subsidized, decreasing costs for fishers. As with the provision of infrastructure, though, the effects of this form of subsidy are likely not large compared to the effects of other types.

Government management support for fisheries may include things that may or may not constitute subsidies, such as fish stocking, releasing

---

64. OCED, *Transition to Responsible Fisheries, Government Financial Transfers.*

65. P. Wallis and O. Flaaten, *Fisheries Management Cost: Concepts and Studies* (Rome: FAO, 2000).

66. Reijinders, "Subsidies and the Environment."

juvenile fish in an area to increase the likely catches, or creating artificial reefs to create safer nurseries for wild fish. Some management services are more obvious subsidies, such as programs designed to assist in marketing fish products or efforts to identify new fisheries to exploit.[67] A few states, such as New Zealand, Iceland, and Australia, have programs to recover costs from fishers for these management services.[68] In most states, however, governments bear the costs.

Decommissioning

If subsidies to encourage increased fishing capacity are problematic, and even subsidies that allow for the efficient operation of fisheries management can cause market distortions that lead to more fishing than would otherwise occur, surely subsidies to encourage fishers to leave the industry should be a better use of resources? Better, perhaps, but these types of subsidies nevertheless cause problems. They include license retirement and boat buyback (decommissioning) programs. By some estimates, these programs constitute the second-largest category of fisheries subsidies within OECD countries.[69]

Buyback programs are becoming increasingly popular as a way to address overcapacity in fisheries. In principle, they are reasonably clear to understand: a government (or on occasion a private) entity offers funding to fishers if they will give up their boats or fishing licenses or both. Such programs simultaneously address the need to reduce fishing capacity and to support those who are no longer able to make a profit from fishing. The best political logic for buyback programs is that they increase the profitability of those who remain in the fishery even if the overall catch needs to be decreased, because there will be fewer vessels to compete for the remaining fish. These programs are also almost always voluntary, so no one is forced to leave the fishery.[70]

---

67. UNEP, *Analyzing the Resource Impact of Fisheries Subsidies*, 21.

68. Ibid., 7.

69. OECD, *Transition to Responsible Fisheries: Economic and Policy Implications.*

70. There are a few exceptions, however. An Australian prawn fishing buyback (and a separate shark fishing buyback) was made mandatory for some fishers after a voluntary program failed to garner enough volunteers. Dan Holland, Eyjolfur Gudmundsson, and John Gates, "Do Fishing Vessel Buyback Programs Work? A Survey of the Evidence," *Marine Policy* 23 (1) (1998): 47–69.

Some consider this form of subsidy to be "good" because it helps to reduce capacity in the fishery.[71] But the effects that these programs have are rarely that clear. By now, dozens of buyback programs have been put in place, with varying goals and mixed levels of success. Even those who generally see the value in these types of programs in theory acknowledge that "in the real world . . . decommissioning schemes are unlikely to help solve the overcapacity problem and could make it worse."[72] Although buyback programs can indeed decrease capacity, at least temporarily, they create a set of perverse incentives that can entice people to enter or remain in the fishery in hopes of future decommissioning incentives. And buyback schemes combined with capital or operational subsidies can lead to a situation of increased capacity.

One of the perverse effects that can come from such programs is an increase in fishing just before buyback programs are put into place. Some buyback programs target the most active fishers in a fishery, generally measured by catch history, in an effort to remove the greatest capacity or pay. One effect of this approach, however, is that it can accelerate overfishing as fishers in areas that expect a capacity-reduction incentive program to be announced increase their fishing behavior in order to inflate their eligibility for a buyback.[73] If a capacity-reduction program is undertaken once, expectations are raised that such a program will happen in the future, and so it can lead to an increase in fishing in anticipation of a future program.[74] Fishers essentially are less afraid of overcapitalization because they believe that in the future the government will be willing to subsidize capacity reduction.[75]

71. Kahn et al., "The Nature and Magnitude of Global Non-fuel Fisheries Subsidies."

72. UNEP, *Analyzing the Resource Impact of Fisheries Subsidies*, 28.

73. Holland, Gudmundsson, and Gates, "Do Fishing Vessel Buyback Programs Work?"

74. Ragnar Arnason, "Fisheries Subsidies, Overcapitalisation, and Economic Losses," in *Overcapacity, Overcapitalisation, and Subsidies in European Fisheries: Proceedings of the First Workshop of the EU Concerted Action on Economics and the Common Fisheries Policy, 28–30 October 1998*, 27–46 (Portsmouth: University of Portsmouth, 1999).

75. A. Cox, *Subsidies and Deep-Sea Fisheries Management: Policy Issues and Challenges* (Paris: OECD, n.d.), http://www.oecd.org/dataoecd/10/27/24320313 .pdf.

A related problem is that remaining vessels or fishers may increase their investment in the fishery. Part of the logic of the programs in the first place is to make fishing more profitable for those who remain in the fishery, but if it becomes more profitable to fish, others will want to enter (unless they are prohibited from doing so). An EU buyback scheme in the 1990s actually increased investment in fishing capacity. Individuals remaining in the fishery when this program operated chose to increase their investments. This new investment was made easier as private creditors offered better terms to fishers for expansion or technology on the assumption that the buyback would increase the profitability of those who remained in the fishery.[76] The periodic availability of such programs also discourages downsizing as fishers hold out for opportunities to be bought out before reducing capacity.[77]

Buyback programs are often structured either intentionally or effectively to target the oldest vessels. There are good reasons to try to remove these vessels from the fishery; they are less safe and may have older technology that makes them more prone to bycatch or ecosystem damage because they can less efficiently target intended stocks. Those who own older vessels are the most likely to sign up for voluntary buyback programs, but these programs almost always end up increasing capacity in the fishery. The vessels that remain are the newest and most technologically sophisticated, and if a small number of new entrants are then allowed, they are virtually guaranteed to be more efficient than those that have been removed.

There is an even bigger problem, however. Unless buyback programs are carefully crafted, they can be subject to the same balloon problems facing fishery management more generally. This is true with respect to whatever it is that the program is buying back. If the program purchases a fishing license, the previous license holder can no longer fish in the fishery targeted for capacity reduction (without obtaining another license, which may even be possible), but that fisher can then take up fishing for a different species or in a different region and with the same

---

76. H. Jorgensen and C. Jensen, "Overcapacity, Subsidies, and Local Stability," in *Overcapacity, Overcapitalization, and Subsidies in European Fisheries*, ed. A. Hatcher and K. Robinson, 239–252 (Portsmouth, UK: Centre for the Economics and Management of Aquatic Resources, University of Portsmouth, 1999).

77. UNEP, *Analyzing the Resource Impact of Fisheries Subsidies*, 25.

vessel and equipment previously used. In some cases, obtaining a new license is not difficult: a buyback program in the New England ground-fish fishery was hampered in its effectiveness by recipients' ability to reenter this same fishery by purchasing previously inactive licenses.[78] Many fishing vessels likewise hold licenses to fish in more than one fishery; the buyback of a license for a given fishery frees up vessels simply to fish elsewhere.[79]

It is rare that the actual boat is purchased in one of these buyback programs, but even when it is, it is likely to show up elsewhere. In an early buyback program in the Canadian salmon fishery in which the government purchased vessels, the vessels were later resold with the condition that they not be used in a Canadian west coast fishery, thus removing the capacity from that area but increasing the likelihood that the vessel would show up in a fishery elsewhere.[80] The EU's Multiannual Guidance Program allows vessels to be exported to non-EU states, and a Norwegian program allowed vessels to be sold to non-Norwegian fishers.[81] Not only are the vessels likely to be transferred, but they are frequently transferred to fisheries that are less well regulated (such as high-seas fisheries) than the ones from which they were decommissioned.[82]

Only programs that completely scrap the purchased vessels are safe from this problem. Even then, however, fishers may use the funds they receive from the program to purchase another vessel or license and fish elsewhere.[83] Ragnar Arnason demonstrates mathematically that not only is the fishing capacity that is removed in this way (via a national

---

78. Theodore Graves and Dale Squires, "Lessons from Fisheries Buybacks," in *Fisheries Buybacks*, ed. Rita Curtis and Dale Squires, 15–53 (Ames, IA: Blackwell, 2007).

79. Ibid., 27.

80. Holland, Gudmundsson, and Gates, "Do Fishing Vessel Buyback Programs Work?" 66.

81. Graves and Squires, "Lessons from Fisheries Buybacks," 33–34.

82. UNEP, Analyzing the Resource Impact of Fisheries Subsidies, 29; OECD, Transition to Responsible Fisheries: Economic and Policy Implications; John Gates, Dan Holland, and Eyjolfur Gudmundsson, *Theory and Practice of Fishing Vessel Buy-Back Program: Subsidies and Depletion of World Fisheries* (Washington, DC: WWF Endangered Seas Campaign, 1997).

83. UNEP, *Analyzing the Resource Impact of Fisheries Subsidies*, 30.

or subnational program) likely to be reallocated elsewhere, but the economic costs of moving the capacity elsewhere will exceed the benefits of the reduction in the original fishery.[84]

Almost all buyback programs thus far have been national or subnational, which then creates international balloon effects when the fishery in question is transnational. An Italian program to buy Mediterranean swordfish gillnet vessels took those ships and fishers out of the fishery but simply allowed for expansion of fishing capacity by other states participating in the fishery.[85]

The most prominent private buyback program is focused on buying the actual boat rather than the rights to fish. The Organization for the Promotion of Responsible Tuna Fisheries, a group of longline tuna fishers in the Indian Ocean, works to buy and scrap secondhand tuna vessels to prevent them from being bought by those who might register them under FOCs and use them to compete in an unregulated way with legitimate tuna fishers.[86] Even this effort, however, has ended up being undermined by nonmember states' expansion in the fishery.[87]

Ironically, then, the type of subsidy that should have the most beneficial impact on reducing capacity generally does not have that effect, in large part because of the global nature of the high-seas fishing industry. The effect of this type of subsidy thus depends dramatically on how the program itself is designed and operated. If both vessels and licenses are bought, the vessel is actually scrapped (rather than resold), and the number of new licenses is severely restricted, such a program (and thus the implicit subsidy that operates it) can have a beneficial effect. It can be particularly effective as a one-time event. But if buybacks are frequent

---

84. Arnason, "Fisheries Subsidies, Overcapitalisation, and Economic Losses."

85. Massimo Spagnolo and Evelina Sabatella, "Driftnets Buy Back Program: A Case of Institutional Failure," in *Fisheries Buybacks*, ed. Curtis and Squires, 145–156.

86. Hiroya Sano, *Are Private Initiatives a Possible Way Forward? Actions Taken by Private Stakeholders to Eliminate IUU Fishing Activities*, AGR/FI/IUU(2004)13 (Paris: OECD, Fisheries Committee, Directorate for Food, Agriculture and Fisheries, 8 April 2004).

87. J. T. Groves and D. Squires, "Requirements and Alternatives for the Limitation of Fishing Capacity in Tuna Purse-Seine Fleets," paper presented to the FAO Methodological Workshop on the Management of Tuna Fishing Capacity, La Jolla, CA, 8–12 May 2006.

enough that they can be anticipated, and if they fail to decrease the number of fishers (and, in particular, the overall capacity—taking account of improved technology), the money that governments spend on these programs may actually increase the number of fishers and their ability to fish, an effect exactly counter to the governments' intentions.

Other Types of Subsidies

Subsidies of other sorts can have important effects on fishing behavior as well. The most obvious are subsidies to the fish-processing industry. The Asia-Pacific Economic Cooperation organization estimates annual fish-processing subsidies in the Asia-Pacific region at roughly US$700 million.[88] This type of subsidization can help encourage fishing by decreasing costs farther along the supply chain, which decreases the costs paid by consumers of fish products and thereby increases demand. Even exchange-rate policy can have a subsidizing effect. For instance, Iceland's 1939 devaluation of its currency was intended in part to increase international demand for Icelandic fish and had the desired effect.[89]

Overseas development assistance, when given to support or encourage the growth of the fishing industry in developing states, can also be considered a subsidy. It tends, however, to work within the subsidy categories discussed earlier. Until the 1990s, this assistance generally went toward capital, technical support, and infrastructure; the amount has more recently declined dramatically, and what remains has tended to go to management efforts.[90] Chapter 7 examines in greater depth the issue of fisheries subsidization as a form of development funding by governments or overseas development agencies. These examples serve to illustrate that governments and intergovernmental organizations can and do intervene in fishers' economic calculations in a variety of ways and for a variety of purposes.

---

88. Asia-Pacific Economic Cooperation, *Study into the Nature and Extent of Subsidies in the Fisheries Sector of APEC Members' Economies* (Singapore: Asia-Pacific Economic Cooperation, 2000).

89. S. Jonsson, *The Development of the Icelandic Fishing Industry 1900–1040 and Its Regional Implications* (Reykjavik: Economic Development Institute, 1981), 186.

90. Jackie Alder, Helen Fox, and Miguel Jorges, "Overseas Development Assistance to Fisheries as a Subsidy," in *Catching More Bait*, ed. Sumaila and Pauly, 54–67.

Conclusions

Subsidies begin with the best of intentions. States want to create economic development for particular regions or sectors or the nation more broadly. They hope to increase the food supply for their populations and encourage full use of resources to which they may have gained legal control. They may want their citizens to be able to compete successfully for the globally unowned supply of high-seas fish.

But subsidies distort the market, removing the price signals it would send to discourage overharvesting of resources. Subsidizing capital and operating costs increases both local and global fishing capacity beyond what local and global stocks can support. Providing price or income supports keeps fishers in a declining industry. Even supporting infrastructure and management practices and buying back excess fishing capacity in an effort to remove its effect on fish stocks can have perverse effects by encouraging more fishing than can be economically and ecologically sustained.

Moreover, subsidies create a political dynamic that leads to their perpetuation. Those who receive them come to depend on them and care deeply for their continuation; the economic pain they will feel if these subsidies are removed is real. In states in which the subsidized population is significant enough to matter regionally but small enough to be supportable without becoming a major national-level economic drain, the political pressure by that population will outweigh the environmental or economic advantages of removing the subsidies. The bottom line, however, is that because the central problem underlying the other regulatory shortcomings facing global fisheries is overcapacity, subsidies need to be removed in order to create a realistic possibility of saving global fisheries.

7

Economic Development

The idea of overseas development assistance as a subsidy brings to the fore the use of fisheries policy as a tool to promote economic development. The use of fisheries as a tool of economic development policy presents a strange irony: governments are at the same time trying to restrict the amount of fish caught and yet are helping to increase the number of people doing the catching. This process is supposed to increase income somehow. The arithmetic of such a proposition is clearly problematic. And yet both governments and international development agencies have subscribed to it, with frequently disastrous results. The idea of fisheries as development policy therefore merits examination as a broader issue than just as a type of subsidy.

The logic driving the use of fisheries resources as a focus of economic development policy is the standard logic of resource extraction and the commons, as noted in chapter 4. Fish in the wild are free in the sense that exploiting them generates economic activity that would not otherwise occur and that does not come at the expense of some other economic activity involving the fish. If they are not fished, then in terms strictly of level of measured economic activity, they are wasted. To the extent that the fish are in a commons, particularly an international commons, if they are not fished by a national industry, they may instead be fished by vessels from elsewhere. In this case, not only are they wasted as a source of current economic activity, but they may be lost as a source of future economic activity as well.

This logic has been at work for centuries. An early example of fisheries policy, for instance, was British subsidization of the Scottish herring fishery in the eighteenth and nineteenth centuries, both through the creation of infrastructure such as port facilities and through the direct payment of bounties per ton of catch. These policies' goals were in part to increase the pool of trained seafarers available to the navy, but also to respond to the domination of the herring fishery by Dutch vessels in English waters in the seventeenth century and to create employment opportunities in one of the poorer parts of Britain.[1] This use of fisheries policy was intended both to take advantage of a free resource to generate employment and to react to the depletion of an accessible commons by a foreign industry. Fisheries have more recently been a focus of development policy in both richer and poorer countries and have been the object of financial and policy support by international development agencies such as the World Bank.[2] Although these organizations are less likely to see fisheries as a tool of economic development than they were in the past, such intentional projects still exist. The logic of fisheries as economic development can still be found in international discourse, particularly with respect to developing countries.

This chapter's core argument is that the logic of using fisheries as a tool of economic development is fundamentally flawed. Over the medium term, underwriting the expansion of a fishing industry will almost always fail as a tool of economic development. It fails in part for the reasons discussed in chapter 3, including the logic of the commons as well as the difficulty that economic (and hence development) analysis has in dealing with natural resources and the value of resources left unused. But beyond these issues, it fails for reasons that are specific to fisheries as the only large industrial-scale exploitation of wildlife.[3] When fisheries development plans come up against the carrying capacity of natural stocks, both ecosystems and industries lose.

When pursuing development, industrialized countries can create fleets capable of fishing out both domestic waters and international commons in a handful of years. Focusing on increasing their capacity as a tool of development causes obvious ecological and ultimately economic problems. Although it may seem surprising, even developing-country fishery expansion and even the promotion of artisanal scale near-shore fishing are

---

1. Callum Roberts, *The Unnatural History of the Sea* (Washington, DC: Island Press, 2007), 120–121.

2. S. M. Garcia, K. Cochrane, G. Van Santen, and F. Christy, "Towards Sustainable Fisheries: A Strategy for FAO and the World Bank," *Oceans and Coastal Management* 42 (1999): 369–398.

3. Wildlife here is distinguished from cultured or raised species.

subject to this problem. Developing countries argue for the right to use fishing as a tool of economic development even while calling on developed countries to stop using this tool.[4] But the flaw in the logic of fishing as a source of economic development applies to developing countries as well, and in most cases a strategy of fisheries promotion will backfire in the medium term. As such, even aside from the argument from a broader fisheries management perspective (developed in chapter 6) that subsidies for industry expansion should be eliminated, developing countries should be wary of expanding small-scale fisheries for domestic political and economic reasons—expansion may well eventually create more problems than it solves.

The use of fisheries resources as an object of economic development policy is counterproductive not only from the perspective of fisheries management, but also from the perspective of sustainable economic development. An examination of the practice of creating or promoting new fishing activity by both governments and international organizations suggests that the practice is self-defeating. Furthermore, arguments that developing countries should be allowed the practice even when developed countries are not should be met with skepticism. There are a few exceptions to this argument, but very few, and we address them in the chapter's final section.

Fisheries as Economic Development

Governments' use of incentives and subsidies to promote the growth of fishing industries as a form of economic development has been practiced for centuries and probably longer.[5] It certainly has taken place since the dawn of industrial fishing, as the British example in the previous section suggests.

In the four decades after World War II, subsidization was used to create large-scale and heavy-industrial fishing fleets in a number of countries.[6] These programs included objectives as varied as the promotion

---

4. Deepali Fernandes, *Running into Troubled Waters: The Fish Trade and Some Implications*, Evian Group Policy Brief (Geneva: Evian Group, November 2006), http://papers.ssrn.com/sol3/papers.cfm?abstract_id=1138271.

5. The practice probably goes back millennia, but older cases are less likely to have been recorded.

6. This practice persisted until it started to become clear that the problem of overexploitation of the world's marine living resources was global rather than local in scale.

of industrialization, employment, and food security. Two examples are illustrative here. In Japan, the motive behind fisheries expansion was expanding the food supply after severe postwar shortages.[7] In the Soviet Union, it was part of a broader policy of industrialization. In both states, fisheries expansion was targeted primarily at international and often distant waters and involved a heavy-industrial model of fishing, with at-sea processing of the catch.[8]

Both fleets therefore ended up targeting species that were previously not heavily exploited. The Soviet fleet in particular made a habit of fishing out species within a few years and moving on to new ones rather than making any real attempt to exploit fisheries sustainably.[9] By the early 1980s, both fleets were huge—Japan's fleet in particular had the world's biggest catch.[10] The Japanese and Soviet programs succeeded in increasing the food supply at a time of particular need, and they developed two of the world's largest fleets of globe-spanning factory fishing vessels.

But in the end neither program worked as an exercise in economic development because neither was sustainable. Much of the Soviet fleet fell victim to the rapid deindustrialization that followed the disintegration of the USSR in 1990;[11] it was no longer economically viable without the support of a centrally planned government. The Japanese government, meanwhile, having invested heavily in building up the industrial fleet and even more heavily in maintaining the industry as it chronically

7. John W. Dower, *Embracing Defeat: Japan in the Wake of WW II* (New York: Norton, 1999), 91–97; G. C. Allen, *Japan's Economic Recovery* (New York: Oxford University Press, 1958), 67–68.

8. Vladimir M. Kaczynski, "Factory Motherships and Fish Carriers," *Journal of Contemporary Business* 10 (1) (1981), 60.

9. Committee on the Bering Sea Ecosystem, Polar Research Board, Commission on Geosciences, Environment, and Resources, and National Research Council, *The Bering Sea Ecosystem* (Washington, DC: National Academy Press, 1996), 169.

10. "Japan's 1982 Fisheries Production Reaches Record High," *Marine Fisheries Review* 45 (7–9) (1983), 30.

11. See, for example, Stale Knudsen and Hege Toje, "Post-Soviet Transformations in Russian and Ukrainian Black Sea Fisheries: Socio-economic Dynamics and Property Relations," *Southeast European and Black Sea Studies* 8 (1) (2008): 17–32.

lost money, has in the past couple of decades invested heavily in shrinking the fleet.[12] In both cases, central goals underlying the construction of distant-water fishing fleets were not met. Neither fleet was economically self-supporting (demonstrated, in the Soviet case, by the implosion of the industry as soon as the subsidy and guaranteed market of centralized planning disappeared). And although the Japanese fleet did create employment and enhance the food supply in the immediate postwar period, the level of subsidy it required may well have generated more sustainable employment if it had been deployed elsewhere in the economy. In a globalized food market, the fact that fish are caught by ships registered in Japan rather than by ships registered elsewhere is largely irrelevant to the security of Japan's food supply.

It is not surprising that investments in deep-water, long-distance fleets intended to capture a resource in a true global commons would quickly generate overcapacity. This result is precisely what one would expect with an international CPR, and the twentieth-century example follows the same path as British subsidies to create a North Sea fleet in the nineteenth century. What is more surprising is that governments do essentially the same thing when the target fishery is in their own waters, to which they should apply a domestic regulatory logic rather than a commons logic. When national waters extended only three miles out to sea, government-driven development of fishing industries as a mechanism for economic development was generally self-limiting in the short term. Local in-shore fish stocks tend to be small and relatively static, meaning that sudden expansion on the sort of scale that would likely result from national policy would quickly and visibly deplete stocks[13] (which is not to say that such expansion was never tried, but rather that it would correct itself relatively quickly).

When national control over fisheries resources was expanded from three miles to encompass two-hundred-mile EEZs, many governments responded first by ejecting foreign fleets and then by focusing on developing domestic capacity to take advantage of their newfound control of some

---

12. Kate Barclay and Sun-Hui Koh, "Neo-liberal Reforms in Japan's Tuna Fisheries? A History of Government–Business Relations in a Food-Producing Sector," *Japan Forum* 20 (2) (2008): 139–170.

13. John Cordell, "Carrying Capacity Analysis of Fixed-Territorial Fishing," *Ethnology* 17 (1) (1978): 1–24.

of the world's richest fishing grounds.[14] To the extent that governments stepped in to help local fishing industries (or create local industries) to fill this vacuum, both aggregate global fishing capacity and the overall amount of pressure on global stocks were likely to increase in response to the experiment in privatization.

The other effect of the increased development of national fishing capacity across states taking advantage of newly nationalized waters was on those fleets that had previously fished the high seas. In some cases, distant-water fleets that no longer had access to stocks in foreign EEZs simply moved closer to home and began fishing in their newly created national EEZs. But there is a mismatch between national fleets and national stocks. The Spanish distant-water fishing fleet, for example, has far more capacity than its local waters can support.[15] There is also a mismatch between the technical demands of fishing close to home and the greater (and sometimes more specialized) capabilities of distant-water fleets.[16]

Distant-water fleets that lost access to their traditional fishing grounds, therefore, often had no choice but to find new stocks to exploit, either by buying access to EEZs or by expanding exploitation of stocks in the global commons.[17] Even if a relatively small proportion of the global distant-water fleet ended up displaced into the newly smaller high seas, the creation of EEZs and the related expansion of national fishing capacities increased the pressure on high-seas stocks.

---

14. The classic example of this process (and one that both responded to early ideas about EEZs and helped set the precedent for broad global acceptance of EEZs) is the Anglo–Icelandic cod war of 1972–1973. See, among other sources, Bruce Mitchell, "Politics, Fish, and International Resource Management: The British–Icelandic Cod War," *Geographical Review* 66 (2) (1976): 127–138, and Mark Kurlansky, *Cod: A Biography of the Fish That Changed the World* (New York: Walker, 1997), 158–169.

15. Ana I. Sinde Cantorna, Isabel Diéguez Castrillón, and Ana Gueimonde Canto, "Spain's Fisheries Sector: From the Birth of Modern Fishing through to the Decade of the Seventies," *Ocean Development & International Law* 38 (2007): 359–374.

16. J. Samuel Barkin and Elizabeth R. DeSombre, "Unilateralism and Multilateralism in International Fisheries Management," *Global Governance* 6 (3) (2000): 339–360.

17. The FAO has recognized this displacement effect as a problem. Its International Plan of Action for the Management of Fishing Capacity (at http://www.fao.org/fishery/ipoa-capacity/legal-text/en) specifically calls on states to act against the possibility. See Steve Cunningham and Dominique Gréboval, *Managing Fishing Capacity: A Review of Policy and Technical Issues*, FAO Fisheries Technical Paper no. 409 (Rome: FAO, 2001).

The Flawed Logic of Fisheries as Economic Development

The use of fisheries development as industrial development policy not only contributes to global overcapacity but is also likely to backfire as economic policy. From a CPR perspective, both the appeal of promoting development of fishing capacity as a form of economic development policy and the problems created by doing so are straightforward. As fishing capacity is increased, the national economy can benefit from a resource that would not otherwise contribute to national economic activity (either because it is left unexploited or because it is exploited by others).[18] The withdrawal of fish from the commons is not counted or experienced (at least in the short term) as a national cost. The problem is the classic commons issue in which individual incentives to overuse a commons undermine the collective interest in sound management, resulting in overcapacity and overexploitation.

This perspective can also explain some of the increase in overcapacity that resulted from the privatization experiment of the creation of EEZs. Although privatization prevents previous users who do not become owners of a commons from continuing patterns of overuse, it also encourages new owners to increase their capacity to use the commons if they did not already have the capacity to fully exploit the resource that they now "own." Unless there is a complementary decrease in capacity by nonowners who previously exploited the resource to balance any increase in capacity by new owners, privatization of a commons can lead to an increase in overall capacity to exploit the resource. It may therefore lead to more pressure on the resource in aggregate.

But while the logic of CPRs suggests that the use of fisheries as development policy can be problematic, it does not get to the core of the problem. Fisheries are different from many other resources in ways that make the investment in or even promotion of the industry as a tool of economic development unlikely to pay out or succeed over the medium term. The dynamics of fish populations generate a fixed upper limit for exploitation both of particular and global stocks (as discussed in chapter 3) that is not to be found in the exploitation of almost any other resource.

---

18. Ragnar Arnason, Kieran Kelleher, and Rolf Willmann, *The Sunken Billions: The Economic Justification for Fisheries Reform* (Washington, DC: World Bank, 2009).

Put simply, fishing beyond MSY on a consistent basis will eventually lead to the decline and possible collapse of a fish stock. There are few other industries for which an analogical maximum possible level of production exists. In most industries, increased capital investment can increase production in the long term as new material inputs to production can be found. There are hypothetical limits on the production of nonrenewable resources. Increasing the amount of oil or copper we produce does hasten the point at which we eventually might run out, but that point nevertheless remains far off in the future.[19] However, as noted in chapter 3, renewable resources are in important ways easier to deplete than nonrenewable resources because production at greater than the rate of replacement undermines renewal. Increased consumption can therefore undermine production potential much more quickly.

This limit to production is in principle true of all renewable resources, but its effect is clearest with fisheries. Production of marine fisheries is already at or beyond its sustainable limit for a majority of major commercial fisheries. And the issue structure of marine fisheries is complicated by both a commons problem and a need for international cooperation. Fishery management provides a contrast, for example, with forestry management, where a smaller proportion of the resource is already being exploited at its maximum potential and where the trees themselves stay in one place and are therefore easier to manage and keep track of.[20] Trees can also be replanted for industrial-scale timber harvesting in a way that open-ocean fish cannot be industrially bred.[21]

This fixed upper limit over the medium term on the amount that can be fished interacts problematically with the logic of economic development.

19. There is even an argument that nonrenewable resources interact with the market such that they will never as a practical matter run out even if they are finite in actual supply. See Julian Simon, *The Ultimate Resource* (Princeton, NJ: Princeton University Press, 1998).

20. For a bleaker assessment of the state of global forestry management, see Peter Dauvergne and Jane Lister, *Timber* (Cambridge: Polity Press, 2011).

21. Aquaculture, the practice of farming fish, functions with a different logic than industrial timber practices. It is much more difficult to control the inputs (and even the breeding) in an ocean environment, and much aquaculture actually uses more fish resources as feed than it produces in farmed fish. See, for example, Paul Molyneaux, *Swimming in Circles: Aquaculture and the End of Wild Oceans* (New York: Thunder's Mouth Press, 2007); Paul Greenberg, "Tuna's End," *New York Times Magazine*, 27 June 2010.

If governments were developing fisheries to maximize food production, then the upper limit would be less problematic. Fishing fleets might in principle be created that have a total capacity of a reasonable precautionary amount below MSY and are maintained at that capacity. But fisheries development as economic development policy runs into two problems, one industrial and one political.

The industrial problem is that economic development strategies are not designed to have steady-state outcomes. The point of development is to grow the economy, and that growth ultimately depends on more efficient use of human and capital resources. More efficient use, in turn, means that given increments of labor and capital produce a greater output over time. In fisheries terms, we can expect the amount of fish that can be caught by a given number of fishers or by a given size and number of fishing vessels to increase over time. This theoretical framework mirrors what has historically happened: fishing efficiency has tended to increase by between 2 and 3 percent per year.[22] The use of fisheries for purposes of economic development is specifically designed to hasten this process by providing capital in the form of vessels and equipment to fishers (or potential fishers). A development policy that does not yield improvements to productivity over time cannot be said to have succeeded.

In practice, this constant tendency to increased efficiency necessarily runs up against the fixed ceiling of available fish. If the economic development strategy creates an industry that can fully utilize available fishery resources or expands an existing industry to that point, then as soon as there is any productivity gain, the amount of labor and/or capital employed by the fishery will need to shrink to avoid overfishing. Even if the industry is intentionally created at a size below what the fishery can support, productivity gains will eventually and necessarily push it up against the limit of what the stock can support. When the industry reaches that point, it will either need to expand to other fishing grounds, thereby exacerbating the global balloon problem, systematically underutilize its capital and labor pools, or actively shrink capital and labor.

---

22. Rögnvaldur Hannesson, "Growth Accounting in a Fishery," *Journal of Environmental Economics and Management* 53 (2007): 364–376; James E. Kirkley, Chris Reid, and Dale Squires, "Productivity Measurement in Fisheries with Undesirable Outputs: A Parametric, Non-stochastic Approach," unpublished manuscript, 12 January 2010.

None of these three possibilities makes for effective economic development policy. Expanding the industry's range puts it into competition with a global fleet that is already over capacity, underutilized, and heavily subsidized, not a particularly promising avenue for development. Subsidizing the formation of capital in an industry that will then necessarily be increasingly underutilized over time is an obviously inefficient use of economic development funds and therefore a questionable target for economic development policy. Building up an industry only to have to decrease its size immediately makes no sense as a form of economic development. To the extent that the purpose of economic development is to increase employment and capital formation, a strategy that not only puts fixed limits on these inputs but also requires that they be immediately and continually scaled back is counterproductive.

Such a strategy is also often politically problematic. Once an economic development policy has generated a fishing industry, it has generated a political constituency for the continued subsidization of that industry. Shrinking the industry requires putting people out of work—people whose political effectiveness as a group may well have been increased because of the development policy. Removing economic support or actively shrinking the industry requires either political confrontation or some form of buyout of existing fishers. In other words, the government would need to pay to eliminate the fishing capacity it has just subsidized the creation of. This pattern can be seen in the Japanese example given earlier and in the Canadian example described later in this chapter.

The alternative to reducing capacity is underutilizing it. This underutilization generally requires strict limits on fishing, which must be implemented in the face of opposition by the political constituency that the development of the industry created. And because fishers have idle capacity, enforcement of these fishing limits is necessary because everyone in the system wants to fish more than is allowed. Compliance with fishing rules is difficult to ensure under the best of circumstances because it can be difficult to determine which vessels are fishing for which species in which manner at any given point in time. Underutilization of national fishing capacity requires expending resources to enforce whatever restrictions are imposed in order to prevent overexploitation in the face of excess capacity.

In such circumstances, limits are likely to be inadequate to the needs of a sustainable fishery. In other words, once the industry has been created as a political constituency, it is likely either to demand subsidization in exchange for the underutilization of capacity or to require additional government funding to reduce capacity. Either way, the use of fisheries resources as a focus for economic development is likely to generate a long-term drain on the national budget that is not taken into account in the creation of the development policy.

Analysts of fisheries policy have recently come to recognize the inherent limits of fishing as a target for economic development policies. A recent World Bank study on development and fisheries, for example, recommends that policy be focused almost exclusively on subsidizing the reduction of capacity in the industry rather than subsidizing or encouraging its creation.[23] Developed countries have for the most part already reached this conclusion themselves. The enthusiasm to expand fishing fleets as part of postwar reconstruction or in response to the creation of EEZs has long since been replaced by efforts to manage capacity and limit capacity utilization. Contemporary examples of developed-country subsidization of expanded capacity, when it happens, tends to be focused on vessels that will be used elsewhere, in the high seas or in other countries' waters, and these efforts are relatively rare.

A Focus on Small-Scale Development?

The most obvious problems of developing fishing capacity as a means of economic development are seen with efforts at developing large-scale industrial fishing operations. It is these operations that can most quickly have global-level impacts on viability of fish stocks. Would a focus on development that prioritizes smaller-scale fishing operations perhaps avoid this counterproductive result?

Experience suggests otherwise. Some governments approached their newly available EEZ resources by self-consciously attempting to encourage smaller-scale, less industrialized exploitation of newly nationalized fisheries resources. The example of Canada and the Grand Banks cod fishery is a good one in this context. The Grand Banks was traditionally one of the world's richest fisheries, one that had supported a large European distant-water fishing fleet for centuries. It was beginning to show some signs of overexploitation (partly as a result of Canadian subsidies

---

23. Arnason, Kelleher, and Willmann, *The Sunken Billions*.

to large-scale industrial operations) when Canada declared its EEZ in 1977, in effect nationalizing a large majority of the fishery. The Canadian government used its newfound authority over the fishery to exclude foreign fleets. Because the Canadian fleet was at that point not big enough to replace the excluded foreign fleets, the fishing capacity at work in the Grand Banks went down significantly. Not surprisingly, fish stocks rebounded. The Canadian government saw this rebound as an opportunity to aid in the economic development of its relatively poor maritime provinces, in particular Newfoundland.[24]

The goal was to encourage a relatively artisanal form of fisheries development to exploit the newly nationalized stock. Government research suggested that the existing fleet tended to be either relatively small vessels in the 30- to 40-foot range or much larger vessels of the 100-foot range. So the new policy split the difference, subsidizing new fishing vessels of up to 65 feet. This resulted in a new generation of vessels of between 64 and 65 feet in length: a policy intended to promote (relatively) artisanal fishing ironically resulted in an increase in the median vessel size.[25] The industry's fishing capacity increased accordingly, faster than had been envisioned. By the mid-1980s, it was clear that fishing capacity exceeded the stock's carrying capacity, but the lesson taken from the immediate post-EEZ period was that the stock recovered quickly, undermining any sense of policy urgency to deal with overfishing.[26]

By the early 1990s, the Grand Banks cod stock had collapsed almost completely and was no longer big enough to support a commercial fishery. A temporary moratorium was put into place, which has since in effect become permanent. The cod stocks have still not recovered, and it is not clear that they ever will.[27] The relatively small vessels of the Newfoundland fleet do not for the most part have enough range to fish far beyond Canada's EEZ, so they do not contribute significantly to global

---

24. For a general overview of this story in the context of the cod fishery more generally, see Kurlansky, *Cod.*

25. Lawrence C. Hamilton and Melissa J. Butler, "Outport Adaptations: Social Indicators through Newfoundland's Cod Crisis," *Human Ecology Review* 8 (2) (2001): 1–11.

26. Kurlansky, *Cod,* 182.

27. Fisheries and Oceans Canada, "What's Holding Back the Cod Recovery?" 2011, http://www.dfo-mpo.gc.ca/science/Publications/article/2006/01-11-2006 -eng.htm.

overcapacity. But, by the same token, their relatively small size means that they cannot be efficiently employed at all. Having invested heavily in building up the fleet in the first place, the Canadian government has spent two decades having to subsidize idle capacity and unemployed fishers.[28]

Developing Countries as a Special Case?

This example demonstrates the counterproductive effects of using fisheries as opportunities for economic development in a wealthy country. But the same logic applies to developing countries as well, and the use of fisheries resources as a target of opportunity for development policy remains more common in the developing world. There even seems to be an emerging consensus in international politics that rules and norms against the subsidization of fishing industry expansion (an essential part of any development policy), even if they were to come into effect, should include exceptions for developing countries. For example, in the current attempt to reduce fisheries subsidies being negotiated in the WTO, the negotiating text calls on developed countries to eliminate subsidies but allows some exceptions for developing countries, and least developed countries are exempted from the rules altogether.[29]

The arguments in favor of differentiated treatment on the question of subsidizing fisheries development are similar to the argument for giving special consideration to developing countries in other international environmental agreements, where the idea is an accepted norm and a common discourse.[30] This approach is understandable. There is, first, the issue of equity. Most of the existing pollution and resource use that international cooperation is designed to combat was done by developed countries, who should therefore bear the bulk of the cost to ameliorate

---

28. See, for example, J. S. Ferris and C. G. Plourde, "Labour Mobility, Season Unemployment Insurance, and the Newfoundland Inshore Fishery," *Canadian Journal of Economics* 15 (3) (1982): 426–441.

29. WTO, Negotiating Group on Rules, *Draft Consolidated Chair Texts of the AD and SCM Agreements*, TN/RL/W/213 (Geneva: WTO, 30 November 2007), annex VIII, "Fisheries Subsidies," 87–90. These negotiations are covered in more detail in chapter 8.

30. For an overview of the concept of common but differentiated responsibilities for developing and developed countries, see Christopher D. Stone, "Common but Differentiated Responsibilities in International Law," *American Journal of International Law* 98 (2) (2004): 276–301.

these environmental problems. Poorer countries, in other words, should share some responsibility for global environmental management but should also be allowed the scope for development that richer countries have already enjoyed. There is also the issue of ability to pay. Developed countries are in a position to bear greater costs than poor countries for the sake of environmental cooperation.[31]

Whatever the extent to which one favors these arguments for differentiated treatment in general, they apply differently, if at all, to fisheries subsidization as a tool of economic development. As a general rule, the logic of differential treatment can interact badly with the logic of the tragedy of the commons. Allowing poor countries a different set of standards works fine as long as those countries do not have the capacity to be major contributors to a commons problem. If they do have the capacity, then differentiated treatment can undermine any benefit gained through cooperation as unregulated or more lightly regulated economic activity grows to replace activity that is more heavily regulated. This problem can be significant in global fisheries, when middle-income countries can easily sponsor major deep-sea fishing fleets[32] and when in any case the ability to reflag vessels in FOCs undermines the idea of country-specific resource exploitation.[33]

The logic of CPRs applies to the problems from expansion of developing country fishing capacity on global fish stocks. The same logic applies to any global commons problem (such as climate change). One can still argue in such a case that if the collective global well-being (or fish supply) decreases but the individual situation of the developing state in question improves, such a trade-off should still be made for the sake of global equity. Why punish developing states for the industrialized world's fisheries overdevelopment, in which they played no role?

---

31. See, for example, Duncan French, "Developing States and International Environmental Law: The Importance of Differentiated Responsibilities," *International and Comparative Law Quarterly* 49 (1) (2000): 35–60.

32. States such as Ghana and Indonesia are in the top thirty registries of ocean-fishing vessel tonnage, with ships actually locally owned. See Institute of Shipping Economics and Logistics, *ISL Shipping Statistics Yearbook 2003* (Bremen, Germany: ISL, 2004), 103.

33. Elizabeth R. DeSombre, "Fishing under Flags of Convenience: Using Market Power to Increase Participation in International Regulation," *Global Environmental Politics* 5 (4) (2005): 73–94.

This logic is based on the idea that even though using the fishing industry as an engine of economic development may in the aggregate create global overcapacity, it remains a viable development strategy. The argument in the previous section suggests, however, that not only is subsidizing the expansion of the fishing industry to exploit underutilized fisheries resources bad environmental policy, it is also likely to be bad development policy and ultimately problematic for the very developing countries it is trying to help. Unless the creation of a fishing industry is done in a way that targets capacity at far less than the target stocks can support in the medium term, it is self-defeating as a development strategy. It will not yield sustainable increases in employment because employment will have to be reduced (or perpetually and increasingly subsidized) as soon as the target fish stock's carrying capacity is reached. It will not yield a growing industry because the industry's size is fixed by the stock's carrying capacity. And it does not represent an efficient use of development capital because as soon as carrying capacity is reached, the capital stock will end up being systematically underutilized.

Building up a fishing industry may be worse economic development policy the poorer the country in question is. Richer countries can better afford to subsidize capacity underutilization or fleet reduction after initial investment in creating excess capacity causes problems. The opportunity cost—spending resources on increased fishing capacity rather than for other development opportunities that are more viable in the long term—for poor countries is greater than for rich countries. And wealthier countries are more likely to have the technical capacity to effectively regulate the fisheries as fishing capacity approaches the stock's carrying capacity.[34]

Differential treatment of developing countries in the context of fisheries development is therefore a bad idea not only from the perspective of global fisheries management, but also from the perspective of national economic development policies. Whereas the ethical basis for a claim to differential treatment may be sound, the economic basis is not. The use of the fishing industry as a vehicle for national economic development is simply too likely to backfire over the medium term and to require governments to invest as heavily in constraining and shrinking the industry as they did in promoting and growing the industry in the first place. Our

34. Wealthy countries have often not availed themselves of the capacity to successfully regulate the fishery in that context, but that is a separate issue.

argument is thus that developing countries should avoid using fisheries development as a tool of economic development not for the good of global fisheries, but for their own good.

International development lending organizations have also come to this conclusion. The World Bank previously attempted to use fisheries resources as a lever for economic development. The Bank's efforts peaked in the 1960s and 1970s and then began to decline as the number of unexploited fisheries available to serve as the targets of development lending decreased. Loans focused on both increasing capacity and infrastructure development as well as on policy development. The Bank's own analysis showed a majority of efforts either failed to generate economic rates of return in the fisheries or failed to increase production altogether.[35]

The World Bank has since backed away almost entirely from lending to projects that increase fishing capacity because it has come to realize that subsidizing new capacity fails not only from a sustainability perspective, but also from a development perspective. It recognizes that it had developed a credibility issue in fisheries and that it needed to focus on providing management expertise rather than capital to increase capacity.[36] Current World Bank efforts in fisheries focus on funding improved management systems and more sustainable fishing techniques as well as on funding capacity reduction and alternative livelihood options among the poorest fishers.[37]

Exceptions?

Should governments always refrain from promoting the expansion of a domestic fishing industry? There are some potential exceptions to the rule, including the development of fisheries that are well under the target fish stocks' carrying capacity, countries with particularly high EEZ-to-landmass ratios, some kinds of artisanal fishing, and recreational fishing.

---

35. Garcia et al., "Towards Sustainable Fisheries."

36. World Bank, *What Is PROFISH? The World Bank's Program on Fisheries* (Washington, DC: World Bank, September 2009), http://siteresources.worldbank.org/EXTARD/Resources/336681-1224775570533/3WhatsPROFISH.pdf.

37. World Bank, Agricultural and Rural Development Department, *Saving Fish and Fishers: Toward Sustainable and Equitable Governance of the Global Fishing Sector*, Report no. 29090-GLB (Washington, DC: World Bank, May 2004).

Policies designed to expand fishing industries need not necessarily be designed to expand them as much as possible. The self-limiting logic of fisheries as economic development applies when fishing capacity approaches the carrying capacity of the target stock. Given constant increases in fishing productivity, policy designed to promote an industry that is able to fully exploit a given stock will quickly create an industry that overexploits it. But if a policy is designed to create an industry with a capacity well under MSY, for example, then the self-limiting aspect will not kick in for a long time. If the policy were to create a fishing capacity only half as big as the carrying capacity of the target stock, for example, then it would be able to continue to operate at its level of employment and capital utilization for a generation before productivity increases would lead to overfishing.

There is in principle no reason why a policy of moderate expansion should not work, although in practice getting the details right can prove difficult, in part because the short-run political and economic gain from further expansion may be difficult to forego in a context where politicians face short time horizons influenced by election cycles.[38] For this approach to be viable, the fish stock in question would have to be excludable in practice (such as a species that exists entirely within an EEZ). Such a policy might be appropriate with fisheries that are not currently being exploited at all. With fisheries that are already exploited, however, there is a possibility that existing capacity in the area, when faced with new entrants to the fishery, would move elsewhere, contributing to the balloon problem. And a policy of moderate expansion would still only delay rather than prevent the limits to the expansion of the industry. As such, this sort of policy is likely more useful as a tool to increase food supplies from a currently unexploited source than it is as an effective tool of economic development.

There are also some specific circumstances, mostly involving small island countries, where it can make sense to promote the development of a domestic fishing industry, although it may well be the case that the development of ancillary industries such as onshore processing contribute more economically than would the development of fishing capacity per se. Classic examples here include the small island states in the Pacific

---

38. Zachary A. Smith, *The Environmental Policy Paradox*, 2nd ed. (Englewood Cliffs, NJ: Prentice Hall, 1995), 46–47.

that control huge EEZs with abundant tuna stocks, such as Kiribati and Tuvalu. For these states, as much as 42 percent of the national economy is generated by the tuna industry, mostly in the form of access fees paid by foreign fleets (or, more likely, their governments) for the right to fish.[39]

In this circumstance, the fishery's carrying capacity is sufficiently large compared with the size of the state's economy that the creation of excess capacity in the medium term is unlikely. Furthermore, the ratio of the size of the fishing industry to the size of the economy is sufficiently large that the governments in question would likely not be able to subsidize the industry meaningfully anyway. Also, the creation of the port and monitoring infrastructure that would be required either for an expanded fishery or for expanded onshore processing facilities would in turn enable better national monitoring of the EEZ, a task for which some of the small Pacific Island states currently have insufficient infrastructure and equipment.[40] The greatest risk of these states' use of fishing resources as an anchor for economic development policy is systemic rather than local. To the extent that the creation of domestic capacity there displaces existing foreign capacity fishing for tuna in these EEZs, it is likely to add to the global balloon problem. And tuna are highly migratory, so increased fishing elsewhere in the ocean will likely affect the availability of stocks even in these areas.

A third category in which the argument against fishing as economic development may not hold in some cases is artisanal fishing, understood as the use of traditional methods or techniques that focus on catching specific fish in small numbers. Most uses of fisheries as the anchor of economic development focus on increasing the level of industrialization of the process by encouraging bigger boats, better engines, more efficient fishing gear, and so on. Policies aimed at replacing industrial with artisanal fishing are in a way an anti-industrial policy, designed to cap or reduce the level of technology in or the level of lethality of fishing. These approaches are thus more likely to be used to expand employment or to

39. Fernandes, *Running into Troubled Waters*, 8.

40. See, for example, Elizabeth Havice and Liam Campling, "Shifting Tides in the Western and Central Pacific Ocean Tuna Fishery: The Political Economy of Regulation and Industry Responses," *Global Environmental Politics* 10 (1) (2010): 89–114. Havice and Campling argue that weak regulatory infrastructure is insufficient by itself to explain existing regulatory patterns, but that it nonetheless plays a significant role.

preserve fishing communities than to generate economic and industrial development. In this respect, they can succeed if used in moderation in the right circumstances.

Two particular circumstances are key. The first is that the target species not be too local. Even if restricted to artisanal techniques, an expanded fishery targeting a purely local stock will find it too easy to fish out the stock. Spearfishing, for example, is likely to do little damage to the stock of a highly migratory species because it is not efficient enough to catch a large proportion of the stock in the small amount of time available as the fish migrate through. It can, however, fish out a small bay or inlet of its sedentary species. The other circumstance is that artisanal fishing (other than subsistence fishing) is most viable when the target species can attract high prices on a market. In the United States, for example, artisanal fishing for species such as lobster in Maine or cod in Massachusetts is viable only because the catch attracts premium prices (and even then only as the number of participants in the industrial part of the fishing process continues to shrink).[41] This observation is more germane the higher local labor costs are: where labor is cheap, artisanal fishing can support fishers even at relatively low market prices.

This viability requires not only demand for the specific species, but often appropriate access to the right markets. Artisanal dayboat cod works in Massachusetts because of proximity to a large and affluent market. It might be more difficult in Alaska, however, although even there an approach to small-scale community development fishing for pollack has been seen as successful.[42] Many potential fisheries in developing countries would require substantial investment in transportation infrastructure in order to function well as economically viable artisanal fisheries. The risk of improving transportation infrastructure in this context is that it would attract more fishers. This risk can be overcome by effective management

41. On lobster, see James M. Acheson, *The Lobster Gangs of Maine* (Lebanon, NH: University Press of New England, 1988). On Massachusetts cod, see Paul Greenberg, *Four Fish: The Future of the Last Wild Food* (New York: Penguin, 2010).

42. J. E. Wilen and E. J. Richardson, "Rent Generation in the Alaskan Pollock Conservation Cooperative," in *Case Studies in Fisheries Self-Governance*, ed. Ralph Edwin Townsend, Ross Shotton, Hirotsugu Uchida, 361–368, FAO Fisheries Technical Paper no. 504 (Rome: FAO, 2008). It should be noted that the quota system used under this program is similar to the ones described in chapter 9.

of the fishery and by mechanisms to exclude newcomers. But in the absence of effective management policies in place, even the development of artisanal fishing can result in the standard pathologies of fishing as economic development.

The final category of exception to the rule against using fisheries resources as a tool of economic development is policy to encourage recreational rather than commercial fishing. Recreational fishing is always likely to present less of a threat to stocks than commercial fishing because the technology it uses is not designed to maximize total biomass catch. One does not, for example, hear about recreational bottom trawling or longlining. The goal of sport fishing is to catch the right fish, not to catch as many fish as possible. Stated differently, recreational fishing results in a much smaller catch per amount of capital and labor deployed than does commercial fishing.[43] At the same time, recreational fishing often generates much more economic activity per amount of fish caught than commercial fishing. For example, although commercial fish landings in the United States are worth about $4 billion per year, the National Oceanic and Atmospheric Administration Fisheries Service estimates that expenditures on recreational fishing equal $31 billion per year.[44] The sport fishery for salmon in British Columbia, although accounting for only 4 percent of the catch, is similarly worth half again as much as the commercial fishery to British Columbia's economy. And each fish caught in the sport fishery is worth twenty-five times as much to the local economy as one caught in the commercial fishery.[45] As a tool of economic development, therefore, recreational fishing, particularly if it is tourism based, can be much more effective than commercial fishing.

The development of recreational fishing has an added benefit over commercial fishing from the perspective of sustainable fisheries management: it can create a political constituency for stricter limits on fishing and fishing practices and for the creation of marine areas that are at least somewhat protected from the most ecologically damaging forms of

---

43. Definitions and regulation of recreational fishing differ from country to country. See the glossary in OECD, *Review of Fisheries in OECD Countries 1997* (Paris: OECD, 1998).

44. National Oceanic and Atmospheric Administration, Fisheries Service, *Recreational Fisheries at NOAA: A Growing Legacy* (Washington, DC: National Marine Fisheries Service, 2011), 3.

45. Scott Steele, "A Haul of Hard Cash," *Maclean's*, 4 August 1997.

commercial fishing.[46] This constituency consists of both the fishers themselves and (particularly in the case of tourism-driven recreational fisheries) providers of services to fishers, ranging from vessels to hotel rooms. Although in many places a recreational fishing constituency is at a disadvantage with respect to commercial fisheries because of the bureaucratic structure of fisheries regulation, the existence of a constituency with a focused economic interest in sustainable management of fishery resources may over time serve as a useful counterweight to regulatory capture.[47]

From a fisheries management perspective, it may well be the case that the creation of recreational fisheries is not just better than commercial fisheries, but a net positive when politics are taken into account. It must be cautioned, however, that this benefit is likely restricted primarily to domestic fisheries. With high-seas or migratory stocks, there is likely to be a spatial mismatch between the recreational and commercial fisheries, and in international politics the latter's representatives can veto the efforts by the former's representatives. In other words, it is likely more difficult for recreational fishers to find an effective political voice at the international than at the national level. It also needs to be noted that recreational fishing needs to be practiced in moderation as well—even the techniques of sportfishing can overexploit a stock if they are used too frequently or by too many people and can have other environmental side effects as well.

Conclusions

The expansion of the fishing industry as a tool of economic development contributes to a global problem of overfishing. It increases capacity in a world in which there is already too much capacity, thereby putting additional pressure on management strategies that are already failing. But it is also the case that expansion of the fishing industry as a tool of economic development is likely to fail in its own terms as well. Note that we are not suggesting that new fisheries should not be exploited or that governments should prevent growth in fishing industries where stocks are underexploited and new capital can yield an economic return without

46. Matt Dozier, "Gone Fishing," *National Marine Sanctuaries*, 12 April 2011, http://sanctuaries.noaa.gov/news/features/0411fishing.html.

47. American Sportfishing Association, "Government Affairs," http://www.asa fishing.org/government/sustainable_fisheries.html.

government assistance. Rather, our argument is that governments should not subsidize expansion in the expectation that it will yield sustainable economic growth.

Fishing is an unusual industry in that there is a fixed upper limit to how big the industry can become over the medium term, and that limit cannot be changed through increased investment or economic effort. Fishing as an industry, in other terms, faces fixed limits to growth shared by few if any other major industries. Subject to the limited exceptions noted earlier, fisheries expansion is unlikely to pay. And it is unlikely to pay whether undertaken by rich countries or by poor countries. Unfortunately, the economic logic that shows that fisheries as economic development policy is in general a bad idea for developing countries has often become lost in developing countries' broader demand in international environmental politics that they receive differentiated treatment. That the claim does not make sense in the specific context of fisheries from a development perspective has gone unexamined in the context of the broader international discourse of environmental sustainability and economic development.

II

Proposed Solutions

8

A Global Fisheries Organization

Current regulatory tools are inadequate to the task of managing the fishing industry. These tools are designed to regulate fishing practice rather than fishing capacity. The concepts underlying regulatory practice are oriented to maximizing yield over the medium term rather than ecological sustainability over the long term. Even if policy were operating in a closed system, without problems of multilevel governance and barriers to exit, uncertainty would undermine the effectiveness of management based on concepts of maximum yield. Although existing management patterns are clearly doing some good in specific cases, they are collectively prone to the balloon problem, in which management rules are arbitraged by fishers, states, and organizations to find uses for excess capacity.

Policy promotes excess capacity even within existing fishing industries both because domestic management structures are often captured by the industries they are supposed to regulate and because those structures are designed primarily to regulate the activity of rather than shrink the size of the existing industry. Regulatory capture means that fisheries policies are designed to maximize the well-being of existing industry members in the short to medium term rather than serve broader long-term social or ecological ends. This regulatory capture contributes to the many and varied subsidies enumerated in chapter 6, allowing the industry to carry much greater capacity than can be supported either by global fish populations or by the market for fish. It is made worse when fisheries capacity is increased in developing countries as part of a policy of economic development, as discussed in chapter 7.

These problems are compounded at the international level. RFMO quotas are consistently laxer than RFMO scientists recommend, as discussed in chapter 4. The existing structure of international fisheries management cooperation, based on RFMOs, regulates entirely on a region- or species-specific basis, making it structurally unable to address the balloon problem even if it can succeed at protecting specific fish stocks in some locations. Reform of this structure that leaves its basic shape intact, however successful it might be in its own terms, cannot overcome this problem. RFMOs and most national-level regulation deal with microlevel solutions, which do not provide the tools for dealing with and can in fact contribute to the macrolevel balloon problem. Furthermore, such successful attempts as have been made at the domestic level to deal with fishing overcapacity cannot translate to the international level through the existing institutional structure because doing so at the regional scale would still suffer from the balloon problem—capacity reduced in one region may simply move to another. It should be stressed here that this is not meant as an argument against either microlevel management or RFMOs. Microlevel regulation is necessary to sustainable global fisheries management, but it is not by itself sufficient.

Existing concepts and structures for regulating global fisheries, then, generally prioritize the fishing industry's needs over the demands of managing and conserving the marine ecosystem (and the interests of society at large in a healthy marine ecosystem) and prioritize microlevel over macrolevel management. These realizations require additional conceptual adjustment. The rest of this book examines how fisheries policies can be reinforced to take seriously societal interests in healthy global fish populations and sustainable management of fisheries on a global scale. Doing so and avoiding the balloon problem require taking the problem of excess fishery capacity seriously. We must cease practices that actively contribute to overcapacity and develop policies that can help to shrink capacity.

Because the problems of capacity, subsidy, and even regulatory decisions (and the related balloon problem they can create) are all collective-action problems at best and tragedies of the commons at worst at the global scale, they must ultimately be solved globally. We therefore make a case for a new international institutional structure. This chapter discusses the way such an institution can deal with global capacity issues. Chapters 9 and 10 present a proposal for a global regulatory approach that would involve the use of tradable quotas to manage—and restrict—global fishing capacity and its access to fish stocks.

Although the creation of a new global institutional structure is not an uncontroversial idea, existing region-based management structures, along with domestic and global sector-based efforts to prevent overcapacity from leading to fishery collapse, have failed. It is time to try something new, building on theoretical analysis and on the successes and failures of previous empirical experience. The suggestions we make in the rest of this book are ambitious—some might argue that they are infeasible. We have two responses to this charge. The first is that our proposals are not "all or nothing." They would be most effective taken together, but implementation of any of the component parts would improve global fisheries management. The second is that although the current political climate may not be good for the sort of institutional change we recommend, it is our hope that making the case for this change will help to change the terms of discourse surrounding the issue and ultimately thereby affect the political climate.

A new global structure might be used to pursue a variety of tasks, from making current patterns of cooperation more efficient to designing new forms of cooperation to address the problem of overcapacity. We begin here by sketching out some ideas on the role of such a structure and its relationship with existing intergovernmental organizations. We then outline some potential tasks, presented in order from the least to the most ambitious. At the former end is coordination of existing fisheries information and management systems. At an intermediate level of ambition, with which we finish this chapter, is an effort to cooperatively reduce subsidies globally. Most ambitious is a global quota system that matches fishing capacity with carrying capacity of stocks, which is the subject of chapters 9 and 10.

Global Institutions and Fisheries Management

A global institutional approach is the best way to bring together what we are calling macroregulation—the regulation of the overall level of fishing capacity in the global system—and microregulation of the way that capacity is used in the actual catching of stocks. The latter, the regulation of specific fishing practices and quantities, is necessarily prone to balloon problems, to regulatory capture, and to the promotion of national fishing industries at the expense of global sustainability. The RFMO system is designed for microregulation and is structurally unsuited to macroregulation. The region- and species-specific focus, combined with the independence of decisions across RFMOs, means that success in one area can result in depletion elsewhere as long as fishing capacity remains the same. Macroregulation is not a substitute for microregulation, which is necessary to sustain the biological viability of specific fish stocks. At the same time, however, RFMOs are not well suited to dealing with the socioeconomic and systemic problem of too much fishing capacity.

The effective macroregulation of global fisheries requires a global regulator in the form of an institution designed to address the many problems that face global fisheries regulation. Such an institution needs to deal with the CPR aspects of global fishing and to develop workable mechanisms for excluding some fishing capacity from the global commons. It must address the problem of industry capture and navigate the complexities introduced by contemporary discourses of fisheries and fisheries management. A global regulator would itself be subject to various challenges from the informational to the political. But it would be better able to face these challenges working in tandem with the RFMO system than leaving the latter to work alone.

In short, then, regulating global fisheries effectively requires an organization that is both global and regulatory. Most RFMOs have the authority to regulate, but none is global. There have been some small moves in this direction. The five RFMOs focusing on tuna and tunalike species—CCSBT, the Inter-American Tropical Tuna Commission (IATTC), ICCAT, IOTC, and WCPFC—have been meeting collectively to try to better organize around common issues, such as scientific research and keeping track of vessels fishing illegally. This process, known as the "Kobe process" after its initial meeting location, aims in principle to "harmonize the activities" of these organizations.[1] On the one hand, that harmonization is not regulatory, and despite collective concerns about capacity by the participating RFMOs, there is no clear way forward on capacity reduction. On the other hand, the formation of a long-term steering committee composed of the leadership of each of the RFMOs[2] is the closest anyone has come to a transregional regulatory process.

---

1. Fisheries and Oceans Canada, "Kobe Process," 2011, http://www.dfo-mpo .gc.ca/international/tuna-thon/Kobe-eng.htm.

2. More information about the Kobe process can be found at http://www.tuna -org.org.

At the same time, although several global intergovernmental organizations deal with fisheries issues, none has the ability to regulate effectively. Most notable is the FAO, which focuses on information gathering and analysis. It collects the most global and authoritative data on fish catches by country and species as well as by weight and market value. It also collects and analyzes data on the importance of fisheries to human diets by country and develops methodologies for assessing the state of marine fish stocks.[3]

The FAO plays two other major roles in global fisheries governance. It provides technical assistance to specific fisheries development projects as well as management and policy assistance to governments in the development of domestic fisheries regulation.[4] And through its Committee on Fisheries, it provides a forum for governments to discuss issues related to fisheries.[5] That committee is at this point the only permanent global intergovernmental forum on fisheries issues. As such, it is often the point of contact for the United Nations' General Assembly regarding fishery-related issues.

Three other global intergovernmental organizations involved in fisheries issues are UNEP, the World Bank, and the WTO. UNEP has several programs focused on marine governance, including the Regional Seas initiative and the marine component of its Ecosystem Management initiative. As is the case with the FAO, these initiatives develop and publicize best practices and provide assistance with specific national programs.[6] The key difference between the two organizations' programs is that UNEP's programs focus on ecosystem management broadly, which encompasses but is not limited to fisheries issues, whereas the FAO's focus is ultimately on the sustainable development of food sources. UNEP also works to increase the visibility of global fisheries issues in the context of the United Nations system and environmental management more broadly.[7]

---

3. FAO, "Statistics—Introduction," http://www.fao.org/fishery/statistics/en.

4. FAO, "FAO Fisheries and Aquaculture Department—Mission," http://www.fao.org/fishery/about/organigram/en#Org-Mission/en.

5. FAO, "Committee on Fisheries (COFI)—Fisheries and Aquaculture Department," http://www.fao.org/fishery/about/cofi/en.

6. P. Akiwumi and T. Melvasalo, "UNEP's Regional Seas Programme: Approach, Experience, and Future Plans," *Marine Policy* 22 (3) (1998): 229–234; UNEP, *Ecosystem Management Programme: A New Approach to Sustainability* (Nairobi: UNEP, 2009).

7. UNEP, *Toward a Green Economy: Pathways to Sustainable Development and Poverty Eradication* (Nairobi: UNEP, 2011).

The World Bank funds and advises specific fisheries-related projects. These projects individually are relatively local in scope, but they are chosen in the context of the Bank's global strategy for fisheries development and governance. As noted in chapter 7, this strategy has come to eschew the creation of new capacity and to promote effective governance, capacity reduction, and the creation of employment alternatives in fishing communities in developing countries. In other words, the strategy does work on a microscale that needs to be done on a macroscale. The World Bank is a development lender, meaning that it has access to quantities of funds orders of magnitude greater than those available to organizations such as the FAO and UNEP, which makes it well placed to deploy carrots with which to encourage developing countries to participate in global fisheries management.

The WTO, meanwhile, has come to play a role in capacity-reduction efforts indirectly through its efforts to reduce subsidization in fisheries. The WTO negotiations, conducted in "rounds," address barriers to free trade globally through negotiated deals among member states. The current Doha Round includes efforts to negotiate reductions of fisheries subsidies.[8] If an agreement through the WTO were reached to decrease (or even prevent) the use of subsidies to the fishing industry, the effect on fishing capacity would be major. And whereas the World Bank can deploy carrots to encourage participation, the WTO can deploy sticks—its rules allow countries to use trade mechanisms to enforce cooperation with multilateral environmental efforts.[9]

These organizations, along with other global and regional intergovernmental organizations, often work together on fisheries-related issues. UNEP's Regional Seas initiative, for example, involves both the FAO and the United Nations Development Programme, and its country-specific initiatives look to the World Bank for funding. But among all this cooperation across organizations, there is no real central coordination of either ideas or effort. And although the various global programs and initiatives provide opportunities and best practices, none undertakes regulation.

---

8. WTO, "Doho Declaration Explained," 2011, http://www.wto.org/english/tratop_e/dda_e/dohaexplained_e.htm.

9. See, for example, Elizabeth R. DeSombre and J. Samuel Barkin, "Turtles and Trade: The WTO's Acceptance of Environmental Trade Restrictions," *Global Environmental Politics* 2 (1) (2002): 12–18.

A Global Fisheries Organization

The FAO, UNEP, the World Bank, and the WTO are global in scale but limited in scope. Some do necessary work on fisheries. The FAO's statistics,[10] for example, are an invaluable resource for understanding global fisheries. Others provide a useful communication function but do not engage in systematic information gathering or actual regulation. The FAO has also tried to change global norms and practices about fishing via the negotiation of nonbinding agreements, including one on fishing capacity,[11] although it is unclear the extent to which these agreements drive rather than reflect changing approaches. Development banks can have significant direct impact on the management of specific fisheries, but exclusively in developing countries and only when those countries choose to seek their input. And the WTO's efforts on fisheries are hostage to the broader deadlock in its Doha Round negotiations.[12] Furthermore, some of the efforts by these organizations overlap. Several of them are involved in collecting statistics of various kinds about global fisheries, and at least three are involved in efforts to address the problem of fisheries subsidies that are not coordinated across organizations.[13] Much of what these organizations do is in fact at the macroscale rather than the microscale, but it is neither authoritative regulation nor centrally organized in any way.

A global fisheries organization would then need to do four things if it is to effectively address the general problem of overexploitation of international fish stocks. It would need either to coordinate or to replace the current functions usefully performed by existing organizations at both the regional and global scales. It would in addition need to be a regulatory body, producing rules that effectively restrict behavior as well as collecting information and determining best practices. And it would need to deal

---

10. See, for example, FAO, *The State of the World's Fisheries and Aquaculture* (Rome: FAO, 2009).

11. FAO, "International Plan of Action for the Management of Fishing Capacity," adopted 1999, http://www.fao.org/fishery/ipoa-capacity/legal-text/en.

12. Florence Chong, "Doha Round Will Fail 'Unless Members Agree,'" *The Australian*, 6 April 2011, http://www.theaustralian.com.au/business/doha-round-will-fail-unless-members-agree/story-e6frg8zx-1226034313883.

13. See, for example, Margaret Young, "Fragmentation or Interaction: The WTO, Fisheries Subsidies, and International Law," *World Trade Review* 8 (2009): 477–515.

with the macroscale, focusing on issues of overcapacity directly rather than on regulating the behavior of an industry whose size is exogenously determined. Finally, it would need to be a fundamentally environment-oriented organization in order to avoid industry capture and to be able to focus on environmental sustainability rather than on the maintenance of existing cultures of fishing.

The first task of a global fisheries organization would be to coordinate or replace other intergovernmental organizations' fisheries-related activities. The question of whether to coordinate or replace would depend on relative institutional efficiency and effectiveness and might vary depending on particular organizations and situations. In some instances, it might make sense for a global fisheries organization to act as an information clearinghouse for data gathered by existing organizations. It would provide meta-analysis of data gathered by RFMOs and by global organizations such as the FAO and WTO. Reducing the number of venues for data gathering risks reducing the amount of data gathered, which is never a good thing. A central authoritative voice in compiling and analyzing those data, however, would reduce some of the uncertainties involved in global fisheries management.

A global fisheries organization would usefully work with other organizations that have existing expertise and resources in the field of development and development assistance rather than attempting to re-create this expertise in-house. For example, the Global Environment Facility (GEF) might be able to provide financial assistance for the capacity-reduction aspect of moving to a system of macroregulation. The GEF exists to provide funding to address the additional costs that developing or economically transitioning countries encounter when they undertake individually costly but collectively advantageous global environmental protection activity.[14] Although the GEF has been historically underfunded, it has already funded some small-scale fisheries projects under its focus on biodiversity conservation and thus has the experience to undertake such a role for a global fisheries organization.

Re-creating the GEF's financial governance mechanisms and creating a new pool of capital to lend would be unnecessarily expensive and would draw a new global fisheries organization's focus away from its core

---

14. Charlotte Streck, "The Global Environment Facility—a Role Model for International Governance?" *Global Environmental Politics* 1 (2) (2001): 71–94.

mission. It would also take a considerable amount of time for the new organization to develop the level of expertise in environmental financing that already exists at the GEF. Similarly, the World Bank's recent efforts at improving fisheries governance and in creating mechanisms to reduce capacity in developing countries[15] need not be replicated in the new organization. A global fisheries organization might most usefully provide goals and general policy guidance to development and financing organizations without trying to replace them. Both the GEF and the World Bank have experience working with global environmental organizations, such as those addressing ozone depletion or climate change, to provide funding for priorities under international agreements.

A global fisheries organization would also not be effective at replacing the microscale regulation currently provided by RFMOs. Although RFMOs face difficulties of regulatory capture and isolation from the global context, the regulatory function they fulfill is necessary, and their expertise in the species and areas they regulate is invaluable. The macroscale regulation we call for in this book is a necessary complement to but not a replacement for microregulation in the form of quotas and rules about fishing behavior and equipment. There are good arguments to be made for providing that microregulation on a region- and species-specific scale. A small scale allows a small group of experts to focus on specific issues, creating knowledge relevant only to a specific set of species and locations. It also helps to prevent specific species or regions in need of microregulation from being forgotten in a focus on larger-scale or more popular fisheries, as would inevitably happen in a single global regulator.

In other words, the advantage of a smaller bureaucratic scale for microscale regulation is that the multiple small organizations retain an incentive to focus on regulating within their specific remit.[16] Furthermore, bringing so many strands of microregulation under the direct authority of just one organization risks the sort of political deadlock currently plaguing the WTO that is preventing any changes in world trade rules.

---

15. Ragnar Arnason, Kieran Kelleher, and Rolf Willmann, *The Sunken Billions: The Economic Justification for Fisheries Reform* (Washington, DC: World Bank, 2009).

16. For a version of this argument with respect to global environmental politics more broadly, see J. Samuel Barkin, "The Environment, Trade, and International Organizations," in *International Handbook of Environmental Politics*, ed. Peter Dauvergne, 334–347 (London: Edward Elgar, 2005).

This sort of deadlock would be disastrous in fisheries regulation, where quotas need to change on an annual basis to reflect changes in stock conditions.[17]

There nevertheless remains a need for a global regulator to coordinate across RFMOs, especially in light of the broader regulatory changes we propose. A global fisheries organization's role in microregulation might range from simple oversight, authoritatively highlighting instances where regulatory capture is having too much impact in determining rules and quotas, to direct bureaucratic involvement. Direct involvement might draw on UNEP's experience in providing secretariat services to existing international environmental agreements,[18] which would provide greater uniformity and transparency across RFMO governance, and a reduction of bureaucratic duplication would provide some cost savings. Such centralization, however, would be at the cost of moving most RFMO secretariats farther away from the fisheries that they regulate and would likely generate political tension. UNEP has found ways around this problem by situating some secretariat functions it produces, such as the Montreal Protocol Multilateral Fund, in purpose-chosen locations. The trade-off would be between local proximity and central coordination.

An effective global fisheries organization should also go beyond coordinating or replacing the roles played by existing global institutions. It should be designed to generate specific rules of behavior rather than focusing exclusively on information gathering and voluntary best practices. International fisheries' CPR structure means that voluntary practices will not suffice for sustainable management because increased overfishing by those who fail to subscribe to the practices can undermine and potentially even eliminate any improvements generated by those who do participate. And although information gathering is necessary to the task of making appropriate fisheries rules and enforcing them, information is not sufficient to change behavior appropriately. By making the collective-action problem inherent in CPR situations clearer or more immediate, better

---

17. Ibid.

18. UNEP provides secretariat services to the Convention on Biological Diversity, the Convention on International Trade in Endangered Species, and the Montreal Protocol on Substances That Deplete the Ozone Layer, among others; this position helps it coordinate across related agreements.

information might even move rational fishers to fish more rather than less in some situations.[19]

A global fisheries organization would of course not be able to legislate in the manner of domestic governments—it would be limited, as are all intergovernmental organizations, to negotiating commitments to rules among only those countries that are willing. It would therefore not be able to address authoritatively the free-rider problem that plagues international cooperation. But those countries willing to be bound by its rules might develop mechanisms of exclusion to discourage noncooperation that would at least help reduce the free-rider problem. Among these mechanisms would be establishing port and market exclusion, keeping fishing vessels flagged in nonmember countries from using port facilities in member countries, and keeping fish caught by vessels flagged in nonmember countries from entry into member countries' markets. Such mechanisms are already incorporated into some RFMO rules.[20] Rules on bycatch or appropriate technology might also be usefully coordinated at this level. A carrot to go along with the stick of mechanisms of exclusion might be transitional funding for developing countries, contingent on participation in global fisheries organization rules.

In order for a global fisheries organization to be effective, its rules would have to focus on regulating at the macroscale as a complement to existing patterns of microregulation provided by RFMOs. The rules must be designed to deal with the aggregate and global problem of overfishing, rather than with particular instances of it. They must, in other words, be designed to deal with the balloon problem. Because this problem ultimately results from too much fishing capacity chasing too few fish, the most straightforward solution to it is to address the excess capacity directly. Decreased capacity would result in fewer free riders and less physical ability to overfish on a global scale, although microregulation would still be necessary to prevent overfishing for specific species or in certain regions.

---

19. L. R. Little, S. Kuikka, A. E. Putnam, F. Pantus, C. R. Davies, and B. D. Mapstone, "Information Flow among Fishing Vessels Modelled Using a Bayesian Network," *Environmental Modeling & Software* 19 (1) (2004): 27–34.

20. Elizabeth R. DeSombre, "Fishing under Flags of Convenience: Using Market Power to Increase Participation in International Regulation," *Global Environmental Politics* 5 (4) (2005): 73–94.

Capacity is easier to police than fishing practice. Hiding fishing behavior that breaks RFMO rules, as noted in chapter 2, is often easy. Excess catch can be kept in hidden holds or transshipped at sea. Illegal fishing gear can be put away when inspectors are near and brought out again when they leave. Excess bycatch or high grading is even easier to hide—the evidence can be thrown out at sea. But excess capacity is harder to hide. Fishing trawlers, in particular those big enough to fish in international waters, are easy to spot and easy to count. A global organization's efforts to coordinate global levels of capacity would help address many of the underlying problems facing fisheries. Many RFMOs already maintain lists of vessels allowed to fish or prohibited from fishing in their areas, and other organizations maintain lists of vessels by registry. It would be possible for a global fisheries organization, as a first step, to track fishing capacity globally.

Changing the Culture of Fisheries Regulation

An effective global fisheries organization would need to be fundamentally an environmental rather than a development organization. It would need to be premised on the recognition that fisheries, beyond a limited artisanal scale, are simply not effective means to sustainable economic development and should therefore not be subject to the same exceptions for developing countries as are many other international agreements. Fisheries are unusual as an international commons resource in which the resource itself is the target of exploitation and in which the limits to exploitation over the long term are rigidly fixed and have already been reached. The relationship between sustainability and development is different in fisheries issues than elsewhere, as explained in chapter 7.

The organization would need to be constituted so as to minimize industry capture. This is a key reason that a global fisheries organization would need to be a new creation rather than a scaling up of or an exercise in cooperation among RFMOs. To the extent that this global organization draws on or remains bureaucratically linked with an existing intergovernmental organization, that organization should be UNEP with its clear environmental focus rather than, for example, the FAO, which focuses on environmental sustainability only insofar as it maximizes the food supply rather than as an end in itself. Drawing on the UNEP model would also involve national environmental bureaucracies in the new organization. Industry representatives and interests presumably would not be excluded altogether (where meaningful regulation is being made, the relevant industrial interests are likely to find a way to have their interests heard) but would at least provide some institutional balance.

A global fisheries organization would also need to institutionalize an environment-centered discourse focused on sustaining the environment rather than on sustaining the fishing industry. This discourse would need to attend first and foremost to the fact that the global fishing industry's current capacity is unsustainable, both ecologically and socially. To focus on the maintenance of the fishing industry or fishing communities in their current form is to argue for maintaining excess capacity and therefore for retaining the balloon problem. Furthermore, as technology gets progressively better, the problem is exacerbated, with more overfishing or fewer fishers or ever greater levels of subsidy. The romantic view of fishers and fishing communities cannot address this pattern adequately. And a focus on local stakeholders in the creation of fisheries regulation gives a veto precisely to those actors whose stakeholding should be the focus rather than the source of regulation.

A focus on capacity in turn requires institutions that rely on the social science of fishing rather than just the natural science. The current RFMO-based institutional structure begins with the natural science of what particular stocks or marine ecosystems can support and then asks what restriction must be placed on fishers in order to keep fishing effort within those limits. Reducing fishing capacity conversely requires regulation that begins with questions about why actors participate in the industry, how to get them to leave it, and how to prevent new actors from joining. This approach will ultimately need to be married to natural science estimates of how much global and regional capacity can be supported by the oceans' marine life. But for the moment overcapacity is sufficiently high that even a successful campaign would likely take decades to shrink fishing capacity to the point where it matched what the oceans can support, particularly because technological improvements will make existing capacity more efficient.

Approaches from economics can be useful in this effort. Despite the limits of economic analysis in the management of fisheries at the microscale discussed in chapter 3 (its biases toward the present over the generational future and toward easily measured utility over things not sold on the market), it has generated some promising policy ideas at the macroscale. At that scale, the problem is the supply of fishers, not the supply of fish. It is, in other words, a social and economic problem that is better suited to a social scientific and econocentric analysis. Environmental economics focuses on creating regulation to ameliorate market imperfections. At the microscale, uncertainties about the capacity of stocks to support exploitation, the transaction costs of monitoring the behavior of fishers, and the property-rights issues of a CPR are difficult to overcome through policy, whether regulatory or market based. At the macroscale, however, market imperfections present less of a problem. The industry is so heavily subsidized and capacity so clearly too high that reducing both is straightforward, given the political will. The market imperfections at the macroscale are also less intractable than at the microscale. Information about fishing capacity is much easier to come by than information about the number of fish. Monitoring the implementation of subsidy and quota-share programs is easier than monitoring the behavior of fishers at sea. And although fish at sea are a CPR, fishers are not, obviating many of the property-rights issues that confront microscale regulation.

The effective macroscale use of new tools that come from environmental economics requires both the political will to refocus regulatory efforts on the supply of fishers before addressing the process of fishing and an interest in crafting new international institutional arrangements to deal with the balloon problem. Doing so is much easier if discussions of fishing include macroscale management, capacity reduction, and sustainable stewardship of the oceans as a whole. The approach needs to emphasize that the fishing industry's current capacity is unsustainable, both ecologically and socially. To focus on the maintenance of the fishing industry or of fishing communities in their current form is to argue for maintaining excess capacity and therefore for retaining the balloon problem and the subsidies required to sustain fishing communities when fish stocks cannot.

Which macroscale ideas that draw from environmental economics and focus on reducing capacity might a new global fisheries organization work to implement? The rest of this book develops two ideas in response to this question, both of which draw heavily on arguments made in earlier chapters. The first is a coordination of the reduction in subsidies to the fishing industry, discussed in the next section. The second—the creation of a system of tradable permits—draws on arguments about CPRs, property rights, and privatization made throughout the first half of this book and is the focus of chapters 9 and 10.

Eliminating Subsidies

One of the most important steps to take to save global fisheries is to cease subsidizing the fishing industry. It is easy to understand why obvious subsidies, such as direct payment for the construction of new vessels, should be eliminated. The broad array of less direct subsidies discussed in chapter 6—including assistance to modernization efforts and the construction and maintenance of port facilities as well as governments' purchase of fishing rights within developing-country EEZs for fishers at no cost to the industry—is also problematic, however. It is one thing to understand the central role that subsidies play in creating the overcapacity problem underlying the fisheries crisis; it is much harder to determine what can or should be done to remedy the problem.

Subsidies can be difficult to remove. The subsidized industry has a concentrated interest in retaining subsidies and likely often has the political power to cause electoral problems for those policymakers who consider subsidy removal. Even if participants in the fishery would benefit collectively if there is less fishing capacity chasing the existing fish, fishers do not want *their* specific benefits to be taken away. In addition, within the domestic contexts of major fishing states, fisheries subsidies are often linked to agricultural subsidies,[21] an even more difficult political mess.

The good news is that there is already evidence that fisheries subsidies are decreasing,[22] and some states or groups of states have intentionally taken actions to diminish the role of subsidies. Important among these actions are the EU's decision in 2002 to reform its Common Fisheries Policy.

---

21. Matteo Milazzo, *Subsidies in World Fisheries: A Reexamination*, World Bank Technical Paper no. 406 (Washington, DC: World Bank, 1998), 18, 25, 33.

22. Ragnar Arnason, "Fisheries Subsidies, Overcapitalisation, and Economic Losses," in *Overcapacity, Overcapitalisation, and Subsidies in European Fisheries, Proceedings of the First Workshop of the EU Concerted Action on Economics and the Common Fisheries Policy (28–30 October 1998)*, 27–46 (Portsmouth: University of Portsmouth, 1999).

The subsidies ended under this revision were those for vessel construction (as of 2004). Other subsidy-ending efforts have been weakened, however; for instance, a 2004 proposal that would have prohibited use of the European Fisheries Fund for all but a few environment- and safety-related forms of equipment or vessel modernization was weakened to allow most vessel modernization as long as it does not increase catch ability.[23]

Development banks and donor countries' enthusiasm for looking at fisheries as a location for economic development has also decreased. The World Bank and other development lenders have switched their focus from helping to further industrialize developing-country fishing operations to funding conservation and management measures.[24]

Although all these developments are positive, they are prone to free-rider (and collective-action) problems. Fisheries are global, but subsidies are domestic. If some states individually decide to eliminate or forego subsidies, thereby decreasing capacity in the global fishery, other states may in response decide to increase their own subsidies in the hopes of attracting a larger share of the global catches. This problem pertains particularly to fishing operations with global range or those that fish for migratory species. It is for this reason that the best hope is for a global cooperative effort to eliminate subsidies, involving some sort of sanction for states that do not participate and continue to subsidize.

Some international efforts have been made to address the collective problem of fisheries subsidies, but so far most of them have failed in some way. Broader efforts (on either trade subsidies or fisheries generally) have dropped reference to fisheries subsidies when they have proved more difficult to address than the negotiations in which they were embedded, or outcomes that did include reference to fisheries subsidies were so watered down as to be meaningless. For example, the Jakarta Mandate on the Convention and Sustainable Use of Marine and Coastal Biological Diversity, a 1995 decision by the Conference of the Parties to the Convention on Biological Diversity, initially considered including fishing subsidies as

---

23. Clare Coffey, "Fisheries Subsidies: Will the EU Turn Its Back on the 2002 Reforms?" WWF, May 2006, http://assets.panda.org/downloads/eu_fisheries_ subsidies_2006.pdf.

24. S. M. Garcia, K. Cochrane, G. Van Santen, and F. Christy, "Towards Sustainable Fisheries: A Strategy for FAO and the World Bank," *Oceans and Coastal Management* 42 (1999): 369–398.

a cause of biodiversity loss that should therefore be avoided. But the issue of subsidies proved "politically sensitive" and was ultimately dropped from the final document.[25]

The OECD's attempts to address fisheries subsidies in shipbuilding failed as well. A binding 1994 Agreement Respecting Normal Competitive Conditions in the Commercial Shipbuilding and Repair Industry prohibited a variety of subsidies for shipbuilding, declaring them "inconsistent with normal competitive conditions in the commercial shipbuilding and repair industry,"[26] but it did not enter into force because the United States refused to ratify it.[27] The agreement, moreover, excluded fishing vessels from its mandate,[28] although these vessels were initially considered within the scope of the negotiations.[29] In 2002, when the OECD concluded that the agreement would not enter into force, it began a new process of negotiation, including not only OECD states, but also those responsible collectively for the vast majority of the world's shipbuilding, including Brazil, China, Croatia, Malta, the Philippines, Romania, Russia, Taiwan, and Ukraine. Negotiations were paused in 2005 when they reached an impasse and were officially abandoned at the end of 2010.[30]

The FAO addresses fisheries subsidies in its International Plan of Action for the Management of Fishing Capacity. This voluntary agreement asks governments to evaluate the contribution of subsidies to overcapacity in fishing, and to "reduce and progressively eliminate all factors, including subsidies" that contribute to excessive fishing capacity.[31] The

---

25. Conference of the Parties to the Convention on Biological Diversity, *Report of the Second Meeting of the Conference of the Parties to the Convention on Biological Diversity*, UNEP/CBD/COP/2/19 (Montreal: CBD, 30 November 1995), 62.

26. OECD, Agreement Respecting Normal Competitive Conditions in the Commercial Shipbuilding and Repair Industry (1994), Annex I.

27. OECD, "Shipbuilding Agreement: Overview," http://www.oecd.org/document/3/0,3343,en_2649_34211_1810179_1_1_1_1,00.html.

28. OECD, Agreement Respecting Normal Competitive Conditions in the Commercial Shipbuilding and Repair Industry (1994), Article 2(2)b.

29. Milazzo, *Subsidies in World Fisheries*, 2.

30. OECD, "Negotiation on a Shipbuilding Agreement," 2011, http://www.oecd.org/document/22/0,3746,en_2649_34211_1823894_1_1_1_1,00.html.

31. FAO, "International Plan of Action for the Management of Fishing Capacity," para. 25 and 26.

International Plan of Action to Prevent, Deter, and Eliminate Illegal, Unreported, and Unregulated Fishing calls on states to refrain from conferring subsidies on "companies, vessels or persons" involved in IUU fishing.[32] These are nonbinding agreements, however, and given the large number of states (as discussed in chapter 6) who still subsidize their fishing vessels, it does not appear to have made any noticeable difference.

The WTO Subsidies Approach

The WTO is concerned with market distortions that have implications for international trade; for that reason, it has recently taken up the issue of fisheries subsidies. The WTO is in theory an excellent venue for an agreement to remove fisheries subsidies: it has global reach, the ability to enforce its agreements, and an increasing concern about environmental issues. The removal of subsidies benefits both the trade context and the environment.

Earlier efforts to address this issue within the WTO failed. The 1994 Uruguay Round Agreement on Agriculture ultimately excluded fisheries products from its scope, although they had been considered in the negotiation process until the very end. Once removed from the agriculture agreement, they then fell under the negotiated Agreement on Subsidies and Countervailing Measures,[33] which prohibits only direct export subsidies and subsidies requiring the use of domestic rather than imported goods, neither of which are prominent categories of fisheries subsidies.

The WTO has made a new effort within the Doha Round to address fisheries subsidies directly. The initial 2001 Doha Ministerial Declaration commits members to "clarify and improve WTO disciplines on fisheries subsidies."[34] The Hong Kong Ministerial Declaration, a 2005 document by WTO member states' trade ministers reaffirming and elaborating on the Doha declaration that began the round of negotiations, includes an agreement to "strengthen disciplines on subsidies in the fisheries sector,

---

32. FAO, "International Plan of Action to Prevent, Deter, and Eliminate Illegal, Unreported, and Unregulated Fishing," www.fao.org/docrep/003/y1224e/y1224e00.HTM, para. 23.

33. WTO, Agreement on Subsidies and Countervailing Measures (1994), part II.

34. Doha WTO Ministerial Declaration, WT/MIN(01)/DEC/1 (November 2001), para. 28.

including through the prohibition of certain forms of fisheries subsidies that contribute to overcapacity and overfishing."[35]

Although there is agreement to address the issue, there are differences of approach. One of the major conceptual disagreements is whether to prohibit subsidies unless explicitly allowed (a top-down approach) or to allow subsidies except for those explicitly prohibited (a bottom-up approach).[36] Not surprisingly, the latter is the approach favored by those (such as the EU and Japan) that most highly subsidize their fisheries.

A 2007 draft negotiation document goes fairly far in prohibiting subsidies, disallowing for developed states all subsidies on capital and operational costs, infrastructure, and price and income support. (Access payments would continue to be accepted, as would subsidies for capacity reduction or vessel decommissioning, as long as all vessels in question are permanently removed from fishing anywhere.) Least developed countries would be allowed to subsidize anything relating to fishing that they chose to, and other developing states would be allowed to subsidize nonmechanized fisheries within their territorial waters and would be able to engage in price and income supports and to subsidize port infrastructure. In addition, capital and operational costs would be allowed to be subsidized in developing states for undecked boats or fishing vessels less than ten meters long.[37]

The WTO negotiations are the most promising attempt to date to deal with fisheries subsidies globally. But the process has major shortcomings. The negotiations take place in the context of the Doha Round of world trade talks, which has been going on for more than a decade and shows no sign of a successful conclusion. Even if WTO member states were to agree to rules to limit subsidies, such an agreement would remain hostage to disagreements on completely unrelated trade issues. Although

---

35. WTO, Hong Kong Ministerial Declaration (22 December 2005), Annex D, para. 9.

36. Ministry for Rural Affairs and the Environment, Department for International Development, *Fisheries Subsidies and the WTO Negotiations*, Policy Brief no. 9 (Valletta, Malta: Ministry for Rural Affairs and the Environment, Department for International Development, 2009), http://www.mrag.co.uk/Documents/PolicyBrief9_Subsidies_insert_Apr09.pdf.

37. WTO, Negotiating Group on Rules, *Draft Consolidated Chair Texts of the AD and SCM Agreements*, TN/RL/W/213 (Geneva: WTO, 30 November 2007), annex VIII, "Fisheries Subsidies," 87–90.

technically possible, it is unlikely that the member states would put a subsidies agreement into effect separately from the broader Doha talks. There is a strong bias in WTO negotiations to keep all issues within comprehensive talks rather than to create separate agreements for separate issues, because separate agreements threaten to unwind complicated networks of concessions on other issues.

Moreover, the subsidies discourse at the WTO is not comprehensive. As noted in chapter 7, even those who agree that subsidies are generally problematic for major commercial fisheries are often willing to make an exception for small-scale fishing operations and "small" vessels. The same kind of exception is frequently sought for developing countries, and these exceptions are in the WTO negotiating texts, allowing for "special and differential treatment"[38] for developing states, especially in terms of small-scale fishing. But the label "small-scale fishing operations" (especially when defined by boat size) says little about the ecosystem effect of these operations, many of which are more efficient and profitable than those using larger vessels and therefore able to harvest more fish per amount of funding.[39]

In this context, a global fisheries organization would play several useful roles. It would act as a coordinator of and clearinghouse for information about subsidies. Different organizations currently collect different information, with different definitions of subsidies and varying levels of completeness. Although WTO member states are required to report their level of fisheries subsidies (within defined categories) to the organization, states underreport this information.[40] Subsidies reported to the OECD or to the Asia-Pacific Economic Cooperation Organization are frequently not reported to the WTO. The OECD in turn collects detailed information about subsidies, but only from its thirty-four members, and even then not all members report information in all categories.[41] There is thus room for an organization that collects data on subsidies globally and proactively,

38. Hong Kong Ministerial Declaration, Annex D, para. 9.

39. WWF, *Small Boats, Big Problems* (Gland, Switzerland: WWF, 2008), http://www.illegal-fishing.info/uploads/wwfsmallboatsbigproblems1.pdf.

40. WWF, *Hard Facts, Hidden Problems: A Review of Current Data on Fisheries Subsidies* (Washington, DC: WWF, 2001).

41. See, for example, OECD, *Review of Fisheries in OECD Countries 2009: Policies and Summary Statistics* (Paris: OECD, 2010).

with one definition of subsidy (or types of subsidy) so that comparisons are possible across states and subsidy types.

The organization would also generate a discussion of subsidies that does not draw on the existing language of special and differentiated treatment. Such language is embedded in many of the organizations currently addressing fisheries subsidies, such as the WTO, in which there has always been exceptions for poorer countries, [42] and the OECD, the membership of which is in practice open only to relatively wealthy economies.[43] A global fisheries organization that begins with the premise that a collapse of global fisheries is a bad thing can address subsidies more comprehensively than existing organizations that begin with a focus on economic development and that see fisheries management only in that context.

Another useful role for a global organization is to centralize discussions and to separate fisheries subsidies from comprehensive agreements (such as trade issues more broadly) that complicate negotiations and can hold an agreement hostage to disagreements on other issues. Having talks managed through a global fisheries organization would increase their profile and would also emphasize that reducing subsidies is ultimately about fisheries sustainability rather than about economic efficiency.

Subsidy-elimination negotiations through a global fisheries organization would still face the difficulties experienced in addressing this topic in other venues. And it is unlikely that a first iteration of an agreement will succeed in eliminating all subsidies right away. A first step, however, would serve as a base onto which further cuts on a broader array of subsidies can be added. By sharpening focus, improving visibility, and highlighting sustainability, a global fisheries organization should make it easier to reach an agreement and to tighten that agreement over time to phase out subsidies of all kinds.

Although a global fisheries organization should be the focus of negotiations in reducing and ultimately eliminating fisheries subsidies, other intergovernmental organizations currently working on the subsidies issue

---

42. For an example of exceptions for poorer countries, see the Agreement Establishing the World Trade Organization (1994), Article XI, para. 2.

43. Member countries need to show policy convergence with a set of OECD standards, which in practice restricts membership to rich and upper-middle-income economies. See OECD, *A General Procedure for Future Accessions*, C(2007)31/ Final (Paris: OECD, 16 May 2007).

would prove useful in providing mechanisms for implementing and enforcing an agreement. In particular, the GEF and the World Bank would
be able to provide both funding and expertise to transition to sustainable
fisheries models in developing countries by improving domestic governance structures and providing funding to transition fishers out of the
industry. And the WTO would be able to provide a mechanism for enforcement. WTO jurisprudence suggests that states can interfere with international trade to further environmental goals, when these efforts are
in support of existing multilateral efforts.[44] Current WTO rules would
therefore allow countries to use trade penalties to retaliate against other
countries that either refuse to participate in subsidies reductions or fail to
live up to their commitments.

Conclusions

We are under no illusion that creating a new global organization and
tasking it—as a first step—with addressing the difficult issues of fishing
subsidies and capacity will be easy. Nor do we provide here a specific
institutional structure for such an organization or a political roadmap for
getting there. But the current state of global fisheries and the inadequacy
of the current, segmented regulatory process demand new approaches to
protecting global fisheries. Therefore, new ideas such as those we propose
need to be put on the table even if the path to realizing those ideas might
not be entirely clear at the outset.

RFMOs, despite their shortcomings, have managed some fish stocks
successfully (especially compared to the situation that would have ensued
without regional management), but the fundamental structure of regulating a global fishing industry by managing stocks regionally is conceptually
flawed and creates the balloon problem described in chapter 3. Regional
management creates too many opportunities for regulatory arbitrage, as
fishers leave jurisdictions with tighter rules for those with looser ones. In
the context of overcapitalized fishing vessels and greater fishing capacity
than existing stocks can support, this regulatory arbitrage both results
from and leads to continued subsidization of the fishing industry.

Breaking this vicious cycle is necessary for realistic management of
existing fish stocks to succeed. Although addressing these problems has

---

44. DeSombre and Barkin, "Turtles and Trade."

been politically difficult thus far, finding ways to manage global fisheries successfully will be a genuine benefit for all those who seek income and sustenance from the sea. It is no longer possible to outcompete others for access to global fish stocks, and the tragedy of the commons that results from efforts to do so confers a short-term cost to states and a long-term loss for human populations and ecosystems. A global organization to manage subsidies and capacity and, as discussed in chapters 9 and 10, to oversee a regulatory system that aligns incentives of fishers with long-term conservation goals may not be created easily, but it is necessary.

9

Individual Transferable Quotas

Although the reduction of subsidies and coordination of general efforts to reduce global fishing capacity are important, these actions alone are not sufficient to save global fisheries. Reducing subsidies will take some air out of the balloon problem but will not impede the flow of the remaining air from one part of the balloon to another. A new global fisheries organization must also be involved in the actual management of fish stocks and of access to them. The approach taken must address the problems of CPRs that have bedeviled existing efforts to protect fisheries. We propose the creation of a globally operated system of tradable permits, often called individual transferable quotas, or ITQs.

This idea draws on analysis about the problems of managing CPRs made in the first part of this book and about the potential advantages of creating property rights as a way around CPR difficulties; it also draws on a significant body of evidence of successful implementation of property-rights based regulation (including of fisheries) at the domestic level. This chapter discusses the environmental economics of privatization as a solution to CPR problems. It also details the conceptual design options of tradable systems of property rights as well as domestic experience with them in fisheries regulation in order to lay the path for a discussion in chapter 10 of how ITQs might be implemented at the international level as a way to regulate global fishing.

Commons Incentive Problems

As the preceding chapters demonstrate, the fundamental structural issue that underpins almost every problem facing global fisheries is the CPR nature of fish as a resource. As with any CPR, fish on the high seas are not excludable: there historically were no rules preventing access, and even when rules limiting access to fisheries resources have been collectively agreed upon, they are in practice extremely difficult to enforce. And because fish are a subtractable resource—fish that are caught become unavailable to others (or for reproduction, to make more fish)—the lack of excludability means that those who do not act to protect the resource may make it impossible for others to do so. Despite some examples of successful small-scale and domestic management of collective fisheries resources,[1] global fisheries have fit the "tragedy of the commons" prediction that "freedom in the commons brings ruin to all."[2] They fit that prediction precisely because of fisheries' CPR incentives.

As suggested in chapter 3, fisheries are an even more complex CPR than other resources, in part because of the nature of the resource itself—fisheries demographics are complex, and populations may be subject to sudden population crashes when numbers are low enough. The interaction of fish with other aspects of their ecosystem also means that the overall amount of a given species fished is not the only variable that determines stock numbers. The health of prey or predator species (which may itself be determined by someone's fishing behavior) and other factors such as pollution and climate change also influence stock size and health. The international nature of ocean fisheries also means that the collective-action problems facing fisheries take the form of a two-level game, in which national regulators face both the international regulatory process and domestic constituencies seeking to avoid or evade regulation.

Many of the problems facing fisheries discussed in the first part of this book can be traced directly or indirectly to the commons nature of fisheries resources. The balloon problem that underpins the shortcomings of regional management of ocean fisheries happens precisely because the commons nature of global fisheries means that people closed out of fishing in one fishery can move to another one. Fishing seems like a reasonable focus for development policy because—in the short term, at least—a state's fishers gaining access to fish they had not previously been able to harvest is like a free source of wealth; they are accessing a shared resource they have not previously accessed.

---

1. Elinor Ostrom, *Governing the Commons: The Evolution of Institutions for Collective Action* (Cambridge: Cambridge University Press, 1990).
2. Garrett Hardin, "The Tragedy of the Commons," *Science* 162 (1968), 1244.

One reason governments of countries at any level of development have subsidized fishing as national policy is to allow their fishers to compete more effectively for the limited supply of fish—they would prefer for their fishers to get a greater share of the collective resource. And some of the overcapitalization of fishing operations, even apart from any government policies, can be traced to individual efforts to gain a greater share of the resource. That incentive may be even more true in a context of limits on catches, and overcapitalization leads in turn to increased pressure for catch limits beyond what the resource can support and to the resulting political capture of the regulatory structure. The unowned nature of global fisheries is thus central to the difficulty in conserving them.

Privatization

One way, therefore, to address the problems that come from the collective nature of fisheries is to find ways to privatize some aspect of them. A private good may still be subtractable, but it *is* excludable. That means that the owner of that good can decide how to use it and will be the only one affected by those decisions. To use Garrett Hardin's terminology,[3] both the positive utility and negative utility that come from using a resource accrue to the same actor. That means the actor is not sharing the negative utility (in Hardin's cow pasture, the finite amount of grass available for grazing) with other potential users of the resource.

The trouble is that ocean fisheries are impossible to privatize as such. In Hardin's metaphoric cow pasture, you can simply divide up the pasture so that each cow herder has a small plot of land into which all that herder's cows have to fit, rather than collectively grazing. A reasonable herder, under those circumstances, would probably not graze more cows than that specific patch could support, but if the herder chose to do so anyway, she would be the only one hurt by her actions; the rest of the herders on privately owned land could continue to graze their cows. There are plenty of situations in which short time horizons might persuade someone to graze more cows than her patch of pasture can support, but in that situation she will be the only one to bear the long-term consequences of overusing the resource.

3. Ibid.

Hardin's metaphor likely understates the extent to which even in the cow pasture example there are negative externalities to overexploiting the resource. Although the cow on one small privatized portion of pasture is not affecting the grass supply for the other cows on other sections as it would have if all cows were grazing on a common pasture, methane produced by the cow's digestion may be affecting the global climate system in a way that affects everyone in the system, or runoff from manure in that one plot may affect a collective water supply. In other words, even those situations that seem ripe for privatization can never fully contain all the effects of individual decisions.

Fisheries present a difficult scenario for privatization. Ownership of particular fish and their offspring is not practical: imagine trying to tag all the fish in the sea and then return them so that the ownership of individual fish caught for sale can be established. You also cannot simply enclose sections of the ocean the way you might enclose portions of a pasture.[4] To the extent that you can attempt to create ownership of particular fishing grounds, doing so may address problems of common access but does not solve all of the negative externalities of overfishing. If one overfishes one's own patch of seafloor, it will likely recover well as long as neighboring patches are not overfished. But this structure presents a classic prisoners' dilemma problem. This kind of area-based privatization is also decreasingly viable the more mobile or migratory the species of fish in question. So if an effort is to be made to change the collective nature of the issue structure for fisheries, some creativity is needed to determine what privatization would even look like for this issue.

This difficulty is further complicated by the two-level aspect of international fisheries regulation—privatization at the domestic level most frequently means giving exclusive rights to individuals,[5] whereas privatization at the international level has traditionally meant giving those rights to states, which is not the same thing and has different implications.

---

4. One exception to the impossibility of enclosure may be the case of aquaculture, but aquaculture is rarely done successfully with highly mobile ocean fish species, and even when it can be done, numerous externalities cannot be contained. For more information on this topic, see Elizabeth R. DeSombre and J. Samuel Barkin, *Fish* (Cambridge: Polity Press, 2011).

5. In some cases, discussed later in this chapter, rights can be given to groups, but that is done less frequently.

Incomplete Privatization—EEZs

The creation of EEZs in 1982, which brought an area two hundred miles from a coastline under state control, was one early effort to undertake this kind of international state-based privatization. It was the single biggest change in global fisheries regulation to date and was driven at least in part by the hope that the creation of excludability in the resource would help with conservation.[6] Some coastal countries had claimed such zones in the 1940s, but they were enshrined in international law in UNCLOS III in the 1980s, and all fishing states have now accepted them. National governments gained complete rights of fisheries management within two hundred nautical miles of shore. EEZs did fundamentally change ocean governance globally, but this move was not as successful as was originally hoped for the conservation of fisheries because of how this privatization was conducted.

Bringing more of the high seas under direct state governance could have addressed the problems of the commons. Most governments, upon creating EEZs, took steps to close off nonnational access to these newly national waters. (As discussed further in chapter 6, they sometimes sell access to foreign fishers, but these deals are individually negotiated exceptions that rely on the area's excludability.) In principle, then, EEZs are efforts to modify the oceans' global commons by making some aspects excludable—states' private property.

The lines drawn in the ocean privatize imperfectly—fish swim in and out of the boundaries drawn as well as between EEZs. These areas have been particularly conflictual,[7] and additional international negotiation was needed to create rules for stocks that "straddle" EEZs or cross between an EEZ and the high seas[8] when it became clear that this incomplete

6. Theodore C. Kariotis, "The Case of a Greek EEZ in the Aegean Sea," *Marine Policy* 14 (1) (1990): 3–14; Robert W. Smith, *Exclusive Economic Zone Claims: An Analysis and Primary Documents* (Dordrecht, Netherlands: Nijhoff, 1986).

7. Edward Miles and William T. Burke, "Pressures on the United Nations Convention on the Law of the Sea of 1982 Arising from New Fisheries Conflicts: The Problem of Straddling Stocks," *Ocean Development and International Law* 20 (4) (1989): 343–357; William T. Burke, "Fishing in the Bering Sea Donut: Straddling Stocks and the New International Law of Fisheries," *Ecology Law Quarterly* 16 (1989): 285–310.

8. Straddling Stocks Agreement (1995).

enclosure would cause problems. But such problems are not the primary reason why this kind of enclosure did not succeed at protecting the fisheries in question.

The management problem comes because in practice EEZs have not privatized the resource in a manner that moved away from CPR incentives. Although states have the right to exclude outsiders from the resources in their EEZs, they often do not limit access by their national fishers. To the contrary, many governments subsidized fishing precisely to expand fishing capacity after EEZ declaration. So although the nationality of those fishing in these areas may have changed, the number of fishers or the amount of fish caught may not have. Over time, in fact, the intensity of fishing in some places even increased, as individual fishers, supported by subsidies, competed with others from the same nationality for fish within newly national waters, in a local replication of what had previously been an international CPR.

Individual Permits

If specific areas of the ocean cannot be directly privatized, then a different way to create property rights is needed. The concept of creating individual rights to things that were previously considered to be collective developed within the realm of environmental politics, beginning with tradable emissions permits for pollutants such as sulfur dioxide and expanding to other types of atmospheric pollutants such as ozone-depleting substances and greenhouse gases as well as to pollution of water. Tradable permits have been created for other previously collective goods, such as access to water itself.[9] The record of success for these mechanisms in some areas has been impressive:[10] for example, the US acid rain program cost 50 percent less than anticipated, yielding annual benefits of $122 billion as of 2010 for a cost of only $3 billion.[11] When the costs of implementing the cap-and-trade program are compared with the estimated costs of the alternative

9. Thomas H. Tietenberg, *Emissions Trading: Principles and Practice*, 2nd ed. (Washington, DC: Resources for the Future, 2006).

10. Jody Freeman and Charles D. Kolstad, eds., *Moving to Markets: Lessons from Twenty Years of Experience* (New York, Oxford University Press, 2007).

11. US Environmental Protection Agency, "Acid Rain Program Benefits Exceed Expectations," n.d., http://www.epa.gov/capandtrade/documents/benefits.pdf.

command-and-control regulations to produce the same emissions reductions, the savings are even more impressive.[12]

An individual allocation of quotas can be helpful for fishery conservation even if those quotas are not transferable. Having a guaranteed allocation of fish helps decrease the "race to fish" and thus the overcapitalization that results, and it can also avoid encouraging dangerous behavior that comes when people are competing with others for the share of a total quota (or even just available amount) of fish. But making that individual allocation transferable increases the economic efficiency of this form of regulation.

The concept behind this general approach is to use the efficiency of the market to create the most cost-effective and politically acceptable way to change behavior. Market mechanisms have long been used to put a price on externalities—or at least to make harmful behavior more expensive and thereby decrease the incentive to engage in it. But although a tool such as a tax can allow for flexibility and entice a change in behavior among those for whom doing so is least costly, the biggest downside to a tax is that it is impossible to predict the total amount of activity (emissions, fishing) that will take place. Conversely, an overall quota (a limit on emissions or on fish caught) can ideally create certainty about the overall amount of activity, but with inefficiency in how that activity and its costs are allocated. Of course, in the case of fishing, as demonstrated in chapter 2, fishing quotas have been problematic as management tools. Setting a quota that is low enough to protect the stock and ensuring that all relevant parties agree to uphold that quota have been unrealistic in a context of overcapacity and regulatory capture. But if quotas can be set well and adhered to, they at least allow for certainty about the overall level of fishing effort.

A tradable-permits system has advantages of both types of regulation. As with a quota, a predictable total amount of the activity in question is allowed. If the quota can be set correctly, there is a defined limit to fish catches. But that limit is divided into portions that are individually owned, so fishers no longer need to race for the catch or take out loans to buy expensive equipment to ensure that they will be able to catch their

12. Dallas Burtraw, *Cost Savings, Market Performance, and Economic Benefits of the U.S. Acid Rain Program*, Discussion Paper no. 98-28-REV (Washington DC: Resources for the Future, September 1998).

share before someone else accesses the fish. The likelihood that under such a system permits will be fished by those who can do so at the least cost (creating a net economic gain in the system) is less important than with emissions reductions, where reducing emissions at the lowest cost makes undertaking them more politically palatable. But enabling lower-cost fishing makes pressure for subsidies or overcapitalization less likely—two factors that have traditionally increased pressure for poor management decisions.

In order for a full property right to be defined as such, the holder of that right needs to be able to make choices about what to do with it—to use it or to sell it to someone else. Systems of tradable permits to both emissions and fish vary in the extent to which they create complete property rights as well as other design features pertaining to allocation of and rules about quotas.

Tradable Permits in Fisheries

Although the concept of tradable permits in fisheries shares some common features with emissions-trading systems, there are also some conceptual differences. The most important difference is how long the quota lasts. With emissions permits, the standard approach is to implement use permits that are used up with the emissions they cover. In other words, if you have a permit to emit one unit of sulfur dioxide, you use up that permit when you emit that unit. Pollution-control tradable-permit systems generally allocate permits anew each year (or each regulatory period) by whatever process of allocation (giving them away, auctioning them) is used.

This process of allocation by regulatory period has several advantages. If permits are sold or auctioned, they can bring in revenue anew in each regulatory period. More important, the number of permits allocated can be changed as environmental or social conditions necessitate. Some systems, such as the US sulfur dioxide emissions-trading system,[13] rely on a regularly scheduled decrease in the number of permits available in order to create a gradual phase down of problematic emissions.

13. James E. McCarthy, *Clean Air Act: A Summary of the Act and Its Major Requirements*, Report for Congress Order Code RL30853 (Washington, DC: Congressional Research Service, Library of Congress, updated 9 May 2005).

In fisheries, as discussed in the next section, the primary approach is to allocate rights to a proportion of the TAC set for a fishery. Although the TAC is set anew for each fishing season, most quota allocations last indefinitely (or at least for multiple years).

The policy also goes by somewhat different names, which sometimes indicate slight differences in approach. There are ITQs, the term and approach most widely used. "Individual fishing quotas," "individual vessel quotas," and "catch shares" are other names for the practice of assigning specific rights to a portion of a catch to specific actors or entities. Although programs by these names may differ with respect to whom the rights are allocated and the extent to which such a right is fully privatized and therefore tradable, the common feature they share is assigning fishing rights.

The biggest advantage of ITQ systems—initially, at least—is the elimination of the overcapitalization that happens in what has been called the "race to fish."[14] If there is a finite amount of fish that can be caught (either because of natural limits or because of collectively adopted quotas), each fisher has an incentive to gain the greatest proportion possible of the available fish because any fish that this particular fisher does not catch can be taken by someone else. This collective structure underpins vessels' increasing technological sophistication (with such fish-locating devices as radar, sonar, geographical positioning systems, and spotter planes, as well as with fish-catching devices such as larger and more technologically sophisticated nets and lines) and their overall size and range. Much of this overcapitalization is also done with the assistance of subsidies (as discussed in chapter 6).

In a system in which fishers are guaranteed access to a portion of the overall quota, however, they no longer need to "race" to ensure that they get the fish they target.[15] As long as a quota has been set at a realistic level, fishers can take their time to get their guaranteed amount and do not need fancier technology or a larger or faster vessel than other fishers to be able to catch their fish. They also do not need to take risks or go fishing in unsafe conditions out of fear of losing their portion of the catch.

---

14. Eugene H. Buck, *Individual Transferable Quotas in Fishery Management*, Report for Congress 95-849 (Washington, DC: Congressional Research Service, Library of Congress, 25 September 1995).

15. T. Essington, "Ecological Indicators Display Reduced Variation in North American Catch Share Fisheries," *Proceedings of the National Academy of Sciences of the United States* 107 (2) (2010): 754–759.

This lack of incentive for overcapitalization not only increases the efficiency (and safety) of fishing in the short run but also addresses the issue of regulatory capture in the longer run. One of the many reasons fishers put pressure on regulators for higher catches than the resource can sustain is their short-run need for income to offset the cost (even when subsidized) of their overcapitalized fishing operations. Although at any given point fishers would rather have more income than less, they will nevertheless be less desperate if they do not have loans on fancy equipment that they need to pay back.

In addition, the subsidization that underlies much of the overcapacity in global fisheries is also much less likely in a system with a guaranteed right to a specific allocation of fish. The incentive to seek or offer subsidies often comes in the context of a need for greater technology to be able to travel farther or faster or find fish more efficiently in order to gain access to catches that others will take if you do not get there first. The "laying up" support or insurance that is a form of subsidy (and that serves to keep fishers in the profession even when market signals suggest they should seek other employment) is less important because a privatized system decreases barriers to exit. If you have a fishing quota and can sell it, you can leave the industry and gain money in the process of doing so.

If such a system would reduce the incentive to create or extend subsidies, it would have broader benefits as well. Subsidies, as elaborated in chapter 6 and elsewhere, create a vicious cycle that makes them extremely difficult to remove once they have been implemented. They create and maintain a constituency that would be worse off if the subsidies were removed. Subsidies thereby increase the number and capacity of fishers, and those fishers advocate strongly to retain the special treatment they receive through subsidies.

Existing Tradable-Permit Systems in Fisheries

The fact that a system of fisheries management should in principle work well does not mean that it will do so in practice. ITQ systems have been implemented in a wide range of fisheries, however, and on average are significantly more effective at managing fisheries sustainably than other systems of regulation. They are far from perfect and still constitute a minority of fisheries regulatory systems globally—as of 2008, only about 10 percent of marine fisheries were regulated using this approach.[16] But there is evidence that they work much better than other systems in wide use, even—or especially—in seriously depleted fisheries.

Tradable-permit systems have been used in a wide variety of fishing contexts over the past four decades. The practice of allocating individual quotas (not initially tradable) began in Iceland in the 1970s. Individual quotas were first made tradable in the 1980s, with programs in Canada and Australia.[17] New Zealand has the most thorough and sophisticated form of ITQ management. This management began in 1983 with deep-water trawl fisheries in the country's EEZ but has since been expanded to cover all fisheries within the EEZ. The protection of ITQs as a permanent property right is enshrined in New Zealand's constitution. Importantly, the initial efforts gave quotas to fixed amounts of catch, but those were later replaced with shares of a future TAC because of the administrative and economic complexity of requiring the government to purchase or sell quotas to achieve the desired catch limits (this system's design features are discussed later).[18]

Nearly three hundred programs now regulate in this manner, covering more than eight hundred species; and more than thirty-five countries have used this fisheries management tool, including 22 percent of the world's coastal countries.[19]

The effect of ITQ systems on fisheries protection has been impressive. A study of fisheries from 1950 to 2003 examined 11,135 fisheries, 121 of which used some form of ITQ system. Controlling for species and region, the study determined that ITQ-regulated fisheries were less than half as likely to collapse than those regulated by other measures. The

16. Cindy Chu, "Thirty Years Later: The Global Growth of ITQs and Their Influence on Stock Status in Marine Fisheries," *Fish and Fisheries* 10 (2) (2008): 217–230.

17. Trevor A. Branch, "The Influence of ITQs on Discarding and Fishing Behavior in Multispecies Fisheries," PhD diss., School of Aquatic and Fishery Sciences, University of Washington, Seattle, 2004.

18. R. Arnason, *A Review of International Experiences with ITQs*, annex to *Future Options for UK Fish Quota Management*, Centre for the Economics and Management of Aquatic Resources (CEMARE) Report no. 58, (Portsmouth, UK: University of Portsmouth, June 2002).

19. Kate Bonzon, Karly McIlwain, C. Kent Strauss, and Tonya Van Leuvan, *Catch Share Design Manual: A Guide for Managers and Fishermen* (Washington, DC:

ITQ-managed fisheries in the study were even more likely than the others to have previously been trending toward collapse, so it appears likely that the ITQ process itself accounts for the dramatic difference.[20]

This process has socioeconomic and regulatory advantages as well. The World Bank and the FAO have concluded that these systems increase economic growth in the countries that practice them,[21] and it has been shown that they enhance fishers' economic well-being (along with job stability and safety)[22] and fishing operations' profitability.[23]

ITQ systems on their own do not address all the problems facing fisheries regulation. One potential problem is the discarding of species caught, either those that are unsought (bycatch) or those that are substandard (high grading). Bycatch discards can in theory be a particular problem for ITQ systems, especially in multispecies fisheries. If fishers are required to hold (or buy) allocations for all fish species they catch, nontarget species are not just unintended, but actively costly, a problem remedied by discarding rather than by landing the unsought species. Whether these fish are thrown back or kept, they are no longer available in the fishery because fish returned to the ocean after being caught rarely survive.[24] Furthermore, fish that are bycatch for some fishing vessels are target species

Environmental Defense Fund, 2010), 3, http://www.edf.org/sites/default/files/catch-share-design-manual.pdf.

20. Christopher Costello, Steven D. Gaines, and John Lynham, "Can Catch Shares Prevent Fisheries Collapse?" *Science* 321 (2008): 1678–1681.

21. Ragnar Arnason, Kieran Kelleher, and Rolf Willmann, *The Sunken Billions: The Economic Justification for Fisheries Reform* (Washington, DC: World Bank, 2008).

22. J. R. Beddington, D. J. Agnew, and C. W. Clark, "Current Problems in the Management of Marine Fisheries," *Science* 316 (2007): 1713–1716; A. Gómez-Lobo, J. Peña-Torres, and P. Barría, *ITQs in Chile: Measuring the Economic Benefits of Reform*, ILADES–Georgetown University Working Papers no. 179 (Washington, DC: School of Economics and Business, Georgetown University, 2007); R. Rose, *Efficiency of Individual Transferable Quotas in Fisheries Management*, Australian Bureau of Agriculture and Resource Economics (ABARE) Report to the Fisheries Resources Research Fund (Canberra: ABARE, September 2002).

23. R. G. Newell, J. N. Sanchirico, and S. Kerr, "Fishing Quota Markets," *Journal of Environmental Economics and Management* 49 (2005): 437–462.

24. Frank Chopin, Yoshihiro Inoue, Y. Matsushita, and Takafumi Arimoto, "Source of Accounted and Unaccounted Fishing Mortality," in *Solving Bycatch: Considerations for Today and Tomorrow*, ed. Alaska Sea Grant College Program, 41–47, Report no. 96–03 (Fairbanks: University of Alaska, 1996).

for others, so discards may affect not only the ecosystem, but other fisheries' profitability.

ITQ systems have at least some advantages for bycatch avoidance. If fishers are required to have quota for all caught fish, the guaranteed access to (target) fish allows fishers to fish more discriminately (rather than in a race for fish) and to use gear and processes designed to access only target fish.[25] There is nevertheless an incentive to discard bycatch that would not result in a profit if a quota needs to be held; such issues will have to be addressed in an ITQ system's design.

Fishers may also still engage in "high grading," a process by which fishers who are allowed to take only a limited amount of fish want to land those of the highest quality or that will fetch the highest price, so they catch more fish than they are allocated, keep the best ones, and throw the less desirable ones overboard.

High grading is actually less likely to be a problem for ITQ-managed fisheries than for those managed by more traditional regulatory approaches, however. In practice, the price differentials across specific fish for the species managed by ITQ systems are not sufficient to motivate high grading.[26] High grading also takes time and resources, both to discard and to catch additional fish,[27] and so the profitability of the higher-quality fish would have to outweigh those costs. The incentive structure created by ownership of the resource also helps decrease the problem of high grading within ITQs. Because fishers do not have to race to ensure their access to the fish, they can catch the fish they seek in a less-rushed process, which may lead to lower discard levels. An examination of existing ITQ systems suggests that most of them do not experience high grading.[28] To the extent that bycatch and high grading discards can still be a potential problem in these systems, some aspects of ITQ design may be able to minimize their likelihood, as discussed in chapter 10.

---

25. R. Arnason, *On Selectivity and Discarding in an ITQ Fishery: Discarding of Catch at Sea*, Working Paper no. 1 (Reykjavik: Department of Economics, University of Iceland, 1996); National Research Council, *Sharing the Fish: Toward a National Policy on Individual Fishing Quotas* (Washington, DC: National Academies Press, 1999).

26. National Research Council, *Sharing the Fish*.

27. M. C. Kingsley, "Food for Thought: ITQs and the Economics of High-Grading," *ICES Journal of Marine Science* 59 (2002): 649.

28. Trevor A. Branch, "How Do Individual Transferable Quotas Affect Marine Ecosystems?" *Fish and Fisheries* 10 (2009): 39–57.

The same problem of leakage from the regulatory area also confronts ITQ systems that face other forms of fishery regulation. If an ITQ system operates in a geographically delineated area, but the fish are capable of leaving the area in question, as is almost always the case, some of the success of a well-managed system can be undermined by the taking of these fish as they leave the area. Not surprisingly, the smaller the area under management, the greater the extent of the problem.[29] ITQs are thus likely to be more successful the larger and more comprehensive they are.

The Design of ITQ Systems

Creators of a tradable-permits system face a set of structural decisions about how it is to be designed. There has been enough history of privatized fishing quotas at the domestic level that both theory and experience can suggest advantages and disadvantages to various design choices. The most common form of this system involves long-term, individually granted rights to a percentage of an individual species' TAC, which is decided at regular intervals. These shares are most frequently allocated by granting rights to existing fishers based on their historical catches, and although the shares are tradable, some restrictions are put in place to avoid concentration of shares.

Each of these design aspects, however, is a choice. Anthony Scott characterizes the conceptual aspects of rights in a fishery as consisting of questions of flexibility, transferability, duration, quality of title, exclusivity, and divisibility.[30] In practical terms, anyone designing an ITQ system must decide what the system grants rights to (catches or capacity or even geographic area), whether species (if catches are the form of regulation) are managed individually or collectively, or whether effort is being managed instead. Whatever right is created, its duration (somewhere between annual and perpetual) must be decided, and the rights must be distributed by some mechanisms across potential rights holders. Assuming a tradable

---

29. Bonzon et al., *Catch Share Design Manual*, 31.

30. Anthony Scott, "Conceptual Origins of Rights-Based Fishing," in *Rights Based Fishing: Proceedings of the NATO Advanced Research Workshop on Scientific Foundations for Rights Based Fishing, 27 June–1 July 1988, in Reykjavik, Iceland*, ed. Philip A. Neher, Ragnar Arnason, and Nina Mollett, 11–38 (Dordrecht: Kluwer Academic, 1989).

right, there might nevertheless be limitations put on trading in order to accomplish some other social goals. Decisions also need to be made about the institutional structure to oversee the trading system and ensure compliance. When considering the possible application of this approach to global management of fisheries, the experience with various aspects of these choices can be instructive with respect to both what makes management sense and what might be politically possible.

Rights to What?

The first choice involves the question of what the property right to be allocated is. It might be access to a portion of the overall TAC (quota), which is the most common approach, used in more than 90 percent of the ITQ systems currently in existence.[31] Or it might be a license to fish in an area that is closed to anyone without a license, or, more likely, it might be a license for a certain tonnage or type of vessel. It might even be a right to fish exclusively in a particular geographic area. This approach is much less common, and it is not entirely clear whether it should even be considered within the category of tradable permits, although it can constitute a form of privatization. At minimum, if fish are not geographically containable, fishing allocation by geography is an incomplete right.

One form of geographically allocated fishery property right is called a territorial use right for fishing (TURF). This approach is most successful for managing species that are mostly sedentary, such as shellfish, or in ocean (or inland water) areas that are fully or mostly enclosed or organized around a particular fixed resource, such as a coral reef or a kelp bed. TURFs give rights to individuals or groups to fish exclusively in a particular geographic area. Some existing TURF programs involve rights that cannot be traded,[32] but in others, such as the Chilean program

---

31. Bonzon et al., *Catch Share Design Manual*, 54.

32. For example, the Maltese dolphinfish fishery allocation is done by nontradable TURF; see Environmental Defense Fund, "Maltese Kannizzati Fishery," Catch Share Design Center, 2011, http://apps.edf.org/catchshares/fishery.cfm?fishery_id=1271. A Swedish system regulating freshwater species and crustaceans also uses this approach; see Environmental Defense Fund, "Swedish Coastal Territorial Use Rights for Fishing (TURF) System," Catch Share Design Center, 2011, http://apps.edf.org/catchshares/fishery.cfm?fishery_id=1327.

regulating access to a wide variety of benthic resources, they are.[33] These programs may sometimes include specific catch limits but conceptually would not require such limits.

One main advantage of this approach is that fishers gain a long-term relationship with a specific local area and may thus be able to gauge their fishing intensity to manage a stock that they will continue to have long-term access to if it is sustainably managed. That is even more true if fishers with TURFs have rights to fish in *only* that one location and cannot simply gain access to other fishing areas if the stocks in that one location disappear. Although TURFs are not widely used and may be difficult to expand to open ocean fisheries, they do operate based on the principle of allocating individual (or collective) rights and may thus provide broader lessons about this approach.

Even among those systems that assign rights to catches, there is a difference between assigning a right to a certain actual amount of catch (usually represented as tonnage) or a percentage of some externally set TAC. The latter is much more commonly used, largely because it allows the actual catch amount to be set differently each year, while the percentage of the overall catch that the right holder has rights to remains constant.[34] A weight-based right, however, can play a similar role if the regulator is willing to play an interventionist role in the market, buying existing quota when the existing allocations exceed the environmentally beneficial level of catch and bringing additional quota onto the market when catches should be allowed to increase. This level of market intervention can be economically or politically difficult, however. New Zealand ITQs were initially a weight-based system but have since been changed to a percentage-based system.[35] And, ultimately (as discussed later), any such system must still include a process for deciding what the target catch should be.

Species Quotas: Single-Species versus Multispecies Systems

Assuming that what is to be allocated is some proportion of the allowable catch of fish, the most common form of property-right system, there is the

---

33. Environmental Defense Fund, "Chilean National Benthic Resources Territorial Use Rights for Fishing Programme," Catch Share Design Center, 2011, http://apps.edf.org/catchshares/fishery.cfm?fishery_id=1802.

34. Bonzon et al., *Catch Share Design Manual*, 58.

35. Arnason, *A Review of International Experiences with ITQs*.

related issue of whether allocations are by single species or by multiple species. The most common form of ITQ management is single species,[36] most likely because the most common form of fisheries regulation allocates quotas to individual species, and it is easier to transition to an ITQ system in the same basic form as the original regulatory process. In an area with only one targeted fish species and little bycatch, single-species management seems to make sense and is certainly easier to operate, but given the importance of bycatch, multispecies management generally makes much more ecological sense.

Multispecies management can be difficult in practice. At minimum, each species' optimum TAC at any given time will be different, so it may not be the case that a quota can be given in total biomass of fish. (Such a system would also increase the incentive to high grade, so that the value of your total landed fish is as high as possible.) Such a "quota basket" is indeed used sometimes, although it was more likely in early ITQ systems and has been abandoned in most of them.[37]

Another way to address multispecies management via ITQs is to give specific quotas for each of a collective set of species but allow weighted transfer of fish from one permit category to another. In other words, if you have a permit for species X but instead catch species Y, you might be able essentially to count a permit for 1.0 unit of species X instead for 0.8 unit of species Y.[38] This approach will work only if there are no especially vulnerable species in the area because an absolute quota on any one species cannot be guaranteed by this method.

Instead of these basket-type approaches, a total quota is usually given simultaneously for multiple species of fish (both target species and bycatch) likely to be landed in a fishery.[39] Because every fish landed needs to have a permit to be legally caught, if fishers are not allowed to trade among species in an allocation or to "bank or borrow" (discussed later), they may have to cease fishing once they reach the cap on any one of the species in their allocation. Tradability of these individual quotas may make it easier for an individual fisher to purchase specific species rights

---

36. Bonzon et al., *Catch Share Design Manual*, 26.

37. Ibid., 28.

38. Ibid.

39. Branch, "The Influence of ITQs on Discarding and Fishing Behavior."

to match what they are catching (or have caught), but it may still be impossible for a given fisher to catch her entire quota across all species if there is no available additional quota to be bought for one or more of the species in the area, especially if fishers are required to stop fishing when they might catch species for which they cannot obtain quotas.[40] There can be benefits to the systematic undercatching of some species quotas that result from this approach; when quotas are unfulfilled, the uncaught fish remain in the ecosystem to reproduce. But concern about an inability to catch a complete quota may lead to some of the "race for the fish" overcapitalization incentives that ITQs can otherwise help avoid.

Monitoring in multispecies ITQ systems is especially important because of the incentive to discard species for which you do not have available quota.[41] Although multispecies management can be difficult both conceptually and practically, it may be necessary for large areas with many species or for ecosystems where one or more species are fished at or beyond capacity.

Effort Management

The other form of right that might be allocated is not to space or amount of catch, but rather to amount of effort. A fisher may be granted a right to a certain number of fishing days or number of lobster traps that can be fished or "effort units" defined some other way. The right might be expressed as a percentage of an overall cap, which can be determined separately and adjusted—in the same way that a TAC can be decided by season—depending on the stock's condition.[42] Australia has used this approach for some lobster and prawn management.[43]

Some RFMOs regulate effort rather than catch. The IATTC has regulated this way historically and has expanded its use of restricting number

---

40. B. R. Turris, "A Comparison of British Columbia's ITQ Fisheries for Groundfish Trawl and Sablefish: Similar Results from Programmes with Differing Objectives, Designs, and Processes," in *Use of Property Rights in Fisheries Management*, ed. R. Shotton, 254–261, FAO Fisheries Technical Paper no. 404/1 (Fremantle, Australia: FAO, 2000).

41. Branch, "The Influence of ITQs on Discarding and Fishing Behavior," 40.

42. Bonzon et al., *Catch Share Design Manual*, 35.

43. Ibid., 152.

and type of vessels rather than catches.[44] The IOTC also began its regulatory efforts by attempting to manage vessel capacity.[45] But it has recently concluded that "limiting fishing capacity, especially when Fleet Development Plans were considered, might not be sufficient to maintain the resource at target levels."[46]

Effort-based rights have some shortcomings. The first is that this approach gives fishers an incentive to innovate to be able to catch more fish with the same equipment or number of days at sea, so it may in some cases lead to precisely the same kind of overcapitalization that property-rights systems are often designed to avoid. As it stands, vessels' fishing efficiency increases by a small percentage annually (in part because of this incentive).[47] There is also a less direct relationship between what you most want to control (the amount of catch) and the mechanism (effort limitation) used. It can be more difficult to predict accurately the relationship between effort (e.g., number of fishing days or number of traps) and overall fish catch than would be the case in a system actually restricting fish catch.[48]

Duration of the Right

A second factor in the creation of a property-rights system is the duration of the right. Unlike emission-based rights, which tend to be single-use

44. FAO, *Managing Fishing Capacity of the World Tuna Fleet*, FAO Fisheries Circular no. 982, FIRM/C982(En) ISSN 0429-9329 (Rome: FAO, 2003), http://www.fao.org/docrep/005/y4499e/y4499e00.htm; IATTC, "Proposal IATTC-82-J-1 Submitted by the European Union: Resolution for the Limitation of Fishing Capacity in Terms of Number of Active Longline Vessels," 82nd Meeting, La Jolla, CA, 4–8 July 2011 http://www.iattc.org/Meetings2011/Jun/PDFfiles/Proposals/IATTC-82-J-1-PROP-EUR-Limiting-longline-capacity.pdf.

45. IOTC, "Collection of Resolutions and Recommendations by the Indian Ocean Tuna Commission," updated April 2010, http://www.iotc.org/English/resolutions.php. See, for example, Resolution 99/01.

46. IOTC, "Approaches to Allocation Criteria in Other Tuna Regional Fishery Management Organizations," prepared by the Secretariat, IOTC-2011-SS4-03[E], 2011, http://www.iotc.org/files/proceedings/2011/tcac/IOTC-2011-SS4-03%5BE%5D.pdf.

47. Rögnvaldur Hannesson, "Growth Accounting in a Fishery," *Journal of Environmental Economics and Management* 53 (2007): 364–376; James E. Kirkley, Chris Reid, and Dale Squires, "Productivity Measurement in Fisheries with Undesirable Outputs: A Parametric, Non-stochastic Approach," unpublished manuscript, 12 January 2010.

48. Bonzon et al., *Catch Share Design Manual*, 151.

permits, rights to an allocation of fish last much longer. To be a true property right, it should be perpetual, which is how New Zealand, Australia, and Iceland allocate shares.[49] The majority of existing ITQ systems allocate their rights perpetually.[50]

Other systems designate the right for a specific period of time, which may be renewable. Shares in the United States are allocated for a decade, with the presumption that they will be renewed.[51] In at least some Canadian ITQ programs, the allocation is only for a year, but quotas are renewed as long as fishers have lived up to their obligations. In practice, then, such systems operate as though they are indefinite, with a built-in compliance mechanism of threatened loss of presumed quota for misbehavior.[52] Danish shares are also indefinite, although they can be revoked after eight years if those who own them are not actively fishing the shares or if they are not otherwise following the rules.[53]

The distinction between a single-use permit and one of long duration is an important conceptual difference between emissions-trading systems and ITQ systems. In an ITQ, what is frequently "owned" is a proportion of a TAC, and the TAC is set based on the fish stock's health. The healthier the stock, the greater the value of your ITQ permit over time. If you own one percent of the overall allowed catch, and the catch allocation increases because of successful management, you are able to take more fish over time by owning the same quota. (The converse is true as well: if the stock becomes depleted, the quota will represent a smaller absolute amount of fish in future years.)

---

49. Ibid., 57.

50. More than 78 percent of species under ITQ management for which an allocation tenure is clearly indicated have perpetual shares. Environmental Defense Fund, "World Catch Share Database," http://www.edf.org/oceans/catch-share-design-center.

51. See the Magnuson–Stevens Fishery Conservation and Management Act, Limited Access Privilege Programs, 16 USC 1853a.

52. D. L. Burke and G. L. Brander, "Canadian Experience with Individual Transferable Quotas," in *Use of Property Rights in Fisheries Management*, ed. Shotton, 151–160.

53. Denmark Ministry of Food, Agriculture, and Fisheries, "Designing an ITQ Management: A Gain for Fishing Communities," 2010, http://www.fvm.dk/Files/Filer/Fiskeri/Danish_Design_of_ITQ.pdf.

By owning a long-term share in the catch, you gain an incentive to think about the fishery's health over the longer term[54] rather than prioritizing short-term gain in the absence of uncertainty about the success of a potential management system. In principle, that stake in the future gives fishers with ITQs an incentive to support both lower catch allocations designed to manage the health of the stock and strong enforcement mechanisms for whatever rules are reached. This incentive exists because the value of their quota, whether they are using it to fish or selling it on the market, will be determined by the fishery's condition.

A related possibility is the idea of temporarily transferring or leasing a share. That approach assumes long-term ownership but gives the owner the opportunity to allow someone else to use the share for a defined period of time. Leasing shares introduces some potential complexities into the system—for instance, the lease price may reflect scarcity of available shares for leasing rather than the fishery's actual economic value[55]—but can be particularly useful in multispecies or bycatch contexts, in which a fisher may need quota for a particular period of time but may not want to hang onto it for the long term.

Allocation Process

Whatever quotas are created within an ITQ system must be allocated in some way. Even before the question of how the allocation is to be distributed, the question of to whom the quota is allocated must be asked. As the term *individual transferable quota* suggests, the vast majority (approximately 90 percent) of quotas are allocated to individual fishers.[56] But these rights can also be allocated to groups or companies or even to communities. One approach used in Canada is to allocate the quota to the vessel itself. Canada also in some cases gives an "enterprise allocation,"

---

54. Anthony T. Charles, "Use Rights and Responsible Fisheries: Limiting Access and Harvesting through Rights-Based Management," in *A Fishery Manager's Guidebook: Management Measures and Their Application*, ed. K. L. Cochrane, 131–157 (Rome: FAO, 2002).

55. M. Gibbs, "The Historical Development of Fisheries in New Zealand with Respect to Sustainable Development Principles," *Electronic Journal of Sustainable Development* 1 (2) (2008): 23–33.

56. Bonzon et al., *Catch Share Design Manual*, 40.

which goes to a fishing company. Although not tradable across companies, this quota may be deployed as the company wishes across its own various fishing vessels.[57]

A truer group allocation of shares gives a fishing quota to a community (which may not be able to transfer the shares outside the community but can allocate access to those shares within the community) or to fishing cooperatives.[58] These shares may sometimes be transferable outside of the community or cooperative, although it is more common that they are not. Group allocations are usually designed to create cooperation within a group and can serve as a form of community development.[59] Despite the existence of such programs, though, the standard form of allocation is still individual. And community allocation schemes do not scale up well to the international level.

The variety of possibilities for how the allocation is distributed is nearly infinite, but there are several common options. An initial decision point is whether the shares will be given or auctioned (sold). Although shares of other tradable rights, such as emissions, are sometimes auctioned, this practice is rare in fish quota programs.[60] Resources allocated by auction have the advantage of bringing in revenue initially, which can help support the regulatory program overall because the operation of such a system can be costly. (If shares are not perpetual, auction-based allocation will bring in revenue more regularly as well.) Auction-based fishing also makes it easier for new entrants to the fishery. Not surprisingly, this approach to allocation is unpopular with the fishers who would have to purchase allocations, especially because prior to the adoption of such a policy, access to fish has been free from their perspective. Allocation by auction is also likely to be more popular with those who have not historically fished in an area (and would thus be less likely to obtain granted quotas, as described later) than with those who have historically fished and thus see access to the fishery as rightfully theirs.

Assuming (as happens most often) that a system grants permits to holders without charging for them, decisions then need to be made about

---

57. Ibid., 41.

58. Ibid., 42.

59. John D. Wingard, "Community Transferable Quotas: Internalizing Externalities and Minimizing Social Impacts of Fisheries Management," *Human Organization* 59 (2000): 48–57.

60. Bonzon et al., *Catch Share Design Manual*, 74.

how to distribute the permits. The most common approach is to base the initial allocation on historical fishing activity. Indicating that the shares are distributed based on historical fishing behavior, however, still requires deciding what behavior, from when, counts toward determining a current allocation.[61] Care needs to be taken in the years before an ITQ system is adopted not to create an incentive in increase catches for the purpose of gaining a greater long-term allocation.[62]

In determining allocations, some systems also take into consideration the level of investment in the fishery (e.g., gear or vessel details) or the extent to which the fisher was involved in developing the initial fishery.[63] It is, in theory, possible to allocate shares equally, but that process rarely happens in practice.[64]

If a tradable-permit system is to work for a subtractable resource such as fish, the allocation needs to cover all actors who might conceivably be fishing—everyone needs to have a share in order to be allowed to harvest fish. Not all systems do allocate fully—it may be that an ITQ is created for a commercial fishery, but not for a subsistence fishery. As has been demonstrated with other fishing rules, however, that division creates an incentive to unregulated fishers to increase their use of technology and the intensity of their fishing effort, especially if a state is using fishing as a development strategy. Unless all potential users of the resource are included in an ITQ system, there is ultimately a serious danger of overfishing.

---

61. D. Huppert, G. M. Ellis, and B. Noble, "Do Permit Prices Reflect the Discounted Value of Fishing? Evidence from Alaska's Commercial Salmon Fisheries," *Canadian Journal of Fisheries and Aquatic Sciences* 53 (4) (1996): 761–768.

62. National Research Council, *Sharing the Fish*; S. Macinko and D. W. Bromley, *Who Owns America's Fisheries?* (Covelo, CA: Center for Resource Economics, 2002).

63. B. Oelofsen and A. Staby, "The Namibian Orange Roughy Fishery: Lessons Learned for Future Management," in *Deep Sea 2003: Conference on the Governance and Management of Deep-Sea Fisheries, Queenstown, New Zealand, 1–5 December 2003, Part 1: Conference Reports*, ed. R. Shotton, 555–559 FAO Proceedings no. 3/1 (Rome: FAO, 2005).

64. In the New Zealand wreckfish ITQ, for example, half the shares were allocated by historical fishing practice, the other half by a system of equal shares. John R. Gauvin, John M. Ward, and Edward. E. Burgess, "Description and Evaluation of the Wreckfish (*Polyprion americanus*) Fishery under Individual Transferable Quotas," *Marine Resource Economics* 9 (1994): 99–118.

Conditions on Trading

In a theoretically complete property right, the holder of that right can do anything with it, without exception. But even with what we usually think of as property, there are usually restrictions about what you can do. Your land may be zoned for a particular use only—allowing you to sell it for residential but not commercial use, for instance—or there may be age restrictions on to whom you can sell your bottle of wine. Likewise, a particular ITQ system may be designed to address certain social priorities, and the market may be created with some restrictions on trading to address these concerns.

If conditions are going to be put on trading, they need to be created as the system is set up. The primary types of conditions involve fleet composition or geography. They are concerned primarily with social issues—the composition of the community of fishers. Some systems are concerned to avoid concentration in ownership of shares in the fishery and so may limit the percentage of permits that can be owned by any given fisher, ranging from a high of 45 percent in the Queensland, New Zealand, system to less than 1.5 percent in some US-based systems.[65] Other systems are concerned with maintaining variety of size or type of fishing operators and so may restrict trades to within size categories or gear type. The Alaska halibut program, for instance, restricts trades to within vessel size and gear categories to ensure that the variety of types of fishing operators continues.[66]

Geographic limitations may serve social functions as well by ensuring that the industry does not concentrate in any one area. Some geographic-based trading restrictions, for instance, are determined by the port at which fish must be landed,[67] but they are also frequently based on ecological conditions, such as ensuring that certain substocks are not overfished.[68]

---

65. Bonzon et al., *Catch Share Design Manual*, 46.

66. C. Pautzke and C. Oliver, *Development of the Individual Fishing Quota Program for Sablefish and Halibut Longline Fisheries Off Alaska* (Anchorage: North Pacific Fishery Management Council, 1997).

67. National Marine Fisheries Service, Alaska Regional Office, "BSAI Crab Rationalization FAQ," 2011, http://www.fakr.noaa.gov/sustainablefisheries/crab/rat/progfaq.htm#qsaifq.

68. Newell, Sanchirico, and Kerr, "Fishing Quota Markets."

Even apart from limits on trades within categories, there may simply be overall limits on number of trades (or on the period during which trades may take place). These limits are most frequently imposed to make it easier to monitor and administratively track permit allocations. Some systems also have restrictions on trading during a transition period, most frequently to allow those involved to learn how the system works.[69]

Details about how the trading process will work must also be decided. They include issues of banking (also called "carryover") and borrowing: Can people save some portion of this year's allocated catch to use next year, or, conversely, can they catch more than this year's allocation with the promise to take it out of next year's catch? Alternately, a period of time at the end of a fishing season may exist during which fishers, in order to cover their catch, can buy the permits they need from those who still have them available.[70] This issue is somewhat more controversial in fisheries quota systems than in emissions systems because the time in which the allocation is used may have a greater effect on the stock's health.[71] Systems that do not allow banking quota for a future year's use can be more successful at rebuilding a stock.[72] In systems with catch shares that are not transferable, however, borrowing serves as a useful compliance mechanism: those who overcatch their limits essentially forfeit the rights to some portion of the following year's catch. This approach is more common than not to allow banking or borrowing, but generally within limits. Most systems that do allow borrowing against a following year's allocation limit the extent to which it can be done or apply some form of discounting so that shares of a future year are actually more costly to use.[73]

---

69. Bonzon et al., *Catch Share Design Manual*, 64–65.

70. Branch, "The Influence of ITQs on Discarding and Fishing Behavior."

71. R. Quentin Grafton, Ragnar Arnason, Trond Bjørndal, David Campbell, Harry F. Campbell, Colin W. Clark, Robin Conner et al, "Incentive-Based Approaches to Sustainable Fisheries," *Canadian Journal of Fisheries and Aquatic Sciences* 63 (3) (2006): 699–710.

72. Branch, "The Influence of ITQs on Discarding and Fishing Behavior"; John H. Annala, Kevin J. Sullivan, and Arthur J. Hore, "Management of Multispecies Fisheries in New Zealand by Individual Transferable Quotas," *ICES Marine Science Symposium* 193 (1991): 321–329; John H. Annala, "New Zealand's ITQ System: Have the First Eight Years Been a Success or a Failure?" *Reviews in Fish Biology and Fisheries* 6 (1996): 43–62.

73. Grafton et al., "Incentive-Based Approaches to Sustainable Fisheries."

Issues about how to deal with new entrants to the fishery also need to be addressed. In principle, ITQs are a market good. But if initial shares are distributed for free based on historical catch (both of which remain the most common distribution rules), those who fished historically gain a right for free that later entrants would have to pay for.[74] Because of this potential inequality between those who had historical access to fishing opportunities and those who would like to have that access, mechanisms are often created to increase access of newcomers in an ITQ system. Some systems, for instance, hold back some of the quota from initial allocation so that it can be distributed or sold (or even leased) in the future to new entrants.[75] Another option is to redistribute existing shares to provide a quota allocation to new entrants. It is much easier to redistribute shares in a system that does not have perpetual ownership of shares, but this latter approach is—not surprisingly—unpopular with those who already hold shares. A slightly less unpopular option involves reverting a small percentage of any traded shares to the management organization for future redistribution. A system that does not allocate long-term shares does not face this problem, but it loses the other advantages of long-term share allocation.

This issue of how to deal with new entrants is likely to be even more contentious in a system operating on the international level, where relative gains are likely to play an even greater role than they do at the domestic level.[76] It is in this context that global inequality is relevant: those who have not had historical access to shares might be further shut out from a system that involves allocation by historical fishing activity and that grants those shares in perpetuity. However, international development organizations might fund the purchase of shares on an existing market, which would allow for increased long-term participation by developing-country fishers without increasing the overall fishing activity.

Other institutional structures are needed as well to create a functioning ITQ system. There must to be a process for managing the trade of permits

74. R. Quentin Grafton, "Implications of Taxing Quota Value in an Individual Transferable Quota Fishery: Comment," *Marine Resource Economics* 11 (1996): 125–127.

75. Bonzon et al., *Catch Share Design Manual*, 48.

76. Robert Powell, "Absolute and Relative Gains in International Relations Theory," *American Political Science Review* 85 (4) (1991): 1303–1320.

and ascertaining whether those engaging in fishing operations or landing fish actually have the right to do so. A system is needed to address non-compliance—to penalize those who fish without the proper permits or who catch more or different types of fish than they are allowed. Some systems allow fish caught beyond allocation to be landed with the payment of a penalty or fee in order to avoid a situation in which fishers discard overages at sea.[77] Setting such a penalty correctly is a challenge; it needs to be low enough to make it worthwhile for fishers with excess catch to report the overage, but high enough that it does not simply become a tax on additional fishing that is worthwhile to pay for the increased number of fish that the fisher can land.[78]

Conclusions

Although it is difficult to create property rights out of something that is a CPR, the existing domestic experience with catch shares in their various forms suggest that this approach can be used with fisheries and can achieve success. These programs, building on the initial success of air pollution regulations in the United States, have adapted the concept to create a fisheries-specific approach that can decrease the incentive for overcapitalization (as well as decrease the incentive for subsidies, increase the opportunities for exit, and reduce the degree of regulatory capture) and give fishers a more direct long-run incentive to conserve fisheries resources.

The most important determinant of the success of an ITQ system may ultimately be the part that is not unique to that management form: the actual setting of quotas. Some regulatory entity must be in charge of aggregating data on catches and stock estimates and determining the overall catch that can be permitted while maintaining (and perhaps even improving) the fish stock's long-term health. If these stock assessments are accurate, the TAC is informed by them, and the level of fishing done is in compliance with them, the fishery's long-term health is likely to improve. Some ITQ systems that have failed to successfully manage the stock failed for precisely this reason: if the allowed catch is too high, it does not

---

77. Branch, "The Influence of ITQs on Discarding and Fishing Behavior."

78. M. P. Sissenwine and P. M. Mace, "ITQs in New Zealand: The Era of Fixed Quota in Perpetuity," *Fishery Bulletin* 90 (1992): 147–160.

matter what kind of system is used to distribute or manage that catch.[79] Compliance with the actual TAC is key as well. In some ITQ *programs*, the problem is not that the TAC was too high, but that it was exceeded.[80] Institutional structures must therefore pay attention to ensuring compliance in ITQs as in any regulatory process.

Any system to protect global fisheries using a property-rights approach must therefore find ways not only to apply existing knowledge about ITQs to the international level, but also to ensure that catch levels and compliance are overseen by those with the knowledge and experience to implement these elements correctly and effectively. Chapter 10 outlines our initial proposal for how a global fisheries organization might act as the coordinating body for such an effort.

---

79. M. Sinclair, R. N. O'Boyle, D. L. Burke, and F. G. Peacock, "Groundfish Management in Transition within the Scotia-Fundy Area of Canada," *ICES Journal of Marine Science* 56 (1999): 1014–1023.

80. Carlos Chávez, Nuria González, and Hugo Salgado, "ITQs under Illegal Fishing: An Application to the Red Shrimp Fishery in Chile," *Marine Policy* 32 (4) (2008): 570–579; Ecotrust, *Catch-22: Conservation, Communities, and The Privatization of B.C. Fisheries: An Economic, Social, and Ecological Impact Study* (Portland, OR: Ecotrust, 2004), http://aquaticcommons.org/1686.

10

International Tradable Permits

Regulations that create something akin to property rights have worked fairly well at regulating fisheries sustainably at the domestic level, working with aspects of local or national legal structures. The challenge of designing an ITQ system for ocean fisheries is that such a system needs to be international and therefore needs to address all of the complications of the two-level system of international fisheries regulation in addition to the complications of creating property rights in a CPR generally. Moving such processes to the international level will be difficult and affects a number of aspects of the design of tradable quota systems.

Several important differences between the domestic and international policy settings must be addressed in the design of any global fisheries ITQ system. The first is the difference between domestic and international law and includes the effects of this difference on the ability to create a comprehensive regulatory structure. International legal obligations are entered into voluntarily. To say that rules are taken on voluntarily does not imply that they are not legally binding, but how one becomes bound and what the consequences are of that binding are different on the domestic level than on the international level. At the international level, in addition to states agreeing to enter into any legal obligations they take on, there are almost always mechanisms to exit these obligations.[1]

The subtractable nature of fish as a resource suggests that effective methods of discouraging free riding are essential because those who do not participate can undermine the conservation undertaken by everyone else. At the same time, the international legal structure makes it difficult to exclude countries from the resource, even if they do not undertake

---

1. Daniel Bodansky, *The Art and Craft of International Environmental Law* (Cambridge, MA: Harvard University Press, 2010).

any regulatory obligations. RFMO management is currently diminished by the presence of fishing vessels flagged in states that are not members of the relevant RFMO; these states (as discussed in chapter 2) may be intentionally choosing to remain outside of the regulatory system so that vessels flagged there will not be legally bound by catch limits or other rules decided by the regulatory bodies.[2] Whereas domestic ITQ systems can be applied authoritatively to all fishing vessels within a national EEZ, an international system would need to overcome the FOC problem.

A second difference that must be addressed is the greater complexity of a global system. Most existing ITQ systems govern only individual species or multiple species within limited geographic areas, whereas an ITQ system that meaningfully prevents fishers from moving to new fishing areas in response to new restrictions where they had been fishing (what we are calling the balloon problem) would need to be comprehensive. Any region or species not covered by the system and not well protected by other regulatory systems such as domestic catch rules within EEZs would become a pressure point in the balloon problem. The regulatory mechanism would therefore need to allow trading across species and regions globally. It would need to be able to oversee a set of relative prices, for example, for quotas of cod and yellowfin tuna rather than just establishing a market price for a single quota. It would also need to allow fishers from, say, China, to trade with fishers from Greenland on an efficient and timely basis.

A third difference is the issue of two-tiered quotas. As discussed in chapter 9, domestic ITQ systems distribute their quotas in various ways, which often involve giving them to existing fishers free of charge as a way to get them to accept the new regulatory system. In international negotiations in the context of RFMO management, states usually negotiate for national quotas that are as large a proportion of TAC as is politically feasible and then distribute the allowed tonnage of a specific catch among members of their national fishing industries as they see fit. A global ITQ system would need to allow for individuals to trade across national lines in a way that undermines the current system of strictly national quotas.

---

2. Elizabeth R. DeSombre, "Fishing under Flags of Convenience: Using Market Power to Increase Participation in International Regulation," *Global Environmental Politics* 5 (4) (2005): 73–94.

International ITQs

In this chapter, we propose a framework for an international ITQ system that takes account of these differences and that can begin both to address the balloon problem and to ameliorate overcapacity in global fisheries. This system would be administered by a global fisheries organization that would work in conjunction with rather than replace RFMOs in the setting of quotas and fishing rules for specific species and regions. What we provide here is only a conceptual framework to make the case for the advantages of such a system; the details would have to be negotiated. The approach laid out here is compatible with a variety of specific structures for such things as the distribution of quotas and the technicalities of trading, so being overly detailed about which form of these structures to choose is unnecessary at this stage. The rest of this chapter discusses the ways in which this framework addresses the differences between the requirements of a domestic and an international ITQ system and why it helps with both the balloon problem and the overcapacity problem more generally.

The basic idea behind a new global ITQ system is to connect the existing patchwork of international regulatory agreements and structures into a more cohesive whole in order to limit as much as possible the scope for arbitrage across and within these structures. Non-quota-based RFMO rules, for example, that prohibit or mandate certain kinds of equipment or close specific fisheries during breeding seasons would be unaffected. RFMO quotas would provide the basis of the new system. RFMOs would continue to generate annual TACs for specific fisheries and species, much as they do now. But once the TACs were generated, they would be part of a global and recurring trading system rather than being distributed anew as national quotas among member states every season.

The system would allow nonstate legal entities (either individuals or incorporated businesses—as discussed further later) to own a permit for a proportion of a given TAC, as is currently standard practice in a number of successful domestic ITQ systems, such as in Iceland, Australia, and Canada.[3] If the TAC were to change in size from year to year, then the

---

3. Ragnar Arnason, *A Review of International Experiences with ITQs*, annex to *Future Options for UK Fish Quota*, Centre for the Economics and Management of Aquatic Resources (CEMARE) Report no. 58 (Portsmouth, UK: University of Portsmouth, June 2002), 26, 8, 14.

amount that can be fished with the permit would change as well. A poorly managed stock might result in collapse of the fishery, which would make the permit worthless, whereas a permit in a well-managed fishery would yield an increasing catch over time (that is, a fixed percentage of an increasing overall TAC). By removing the annual process of distributing quotas among member countries, the system would act to break the current connection in RFMO negotiations between overall TAC and national quota and would therefore make the negotiation of the TAC less political.

Removing this annual process would also decrease states' propensity to submit formal objections and thus exempt themselves from RFMOs' otherwise binding provisions. Some of the most contentious moments of opting out of rules, such as the EU's objection to the turbot quotas within NAFO in 1995,[4] have not been about the TAC itself, but about its allocation. If states within RFMOs are able to agree on a TAC without having also to agree on how it will be allocated across states (because a percentage of the TAC is both already owned and tradable), it removes one of the barriers to adopting reasonable catch limits.

Long-term ownership of quota shares is one important innovation. A second is tradability—quotas can be sold and bought. In effect, quota shares can be traded across species and even across regions. In practice, such a system would not likely involve actual trading of a specific permit from one region for a permit from another. But the fact that permits can be bought and sold means that in effect they would be tradable globally (although presumably only among nationals of states that are members of the system). Fishers displaced by falling TACs in one fishery can therefore move to another (as long as they have purchased the appropriate quota) but still stay within an overall management system. Fishers would thus gain a long-term commitment to the health of the particular fishery in which they hold permits because the value of those permits depends on the fishery's health. But at the same time they would have less fear of lower TACs preventing them from fishing in the short term because they would be able to buy access to fisheries elsewhere.

The key to such a system's success is comprehensiveness. It must encompass all international fisheries, meaning those that cannot be

---

4. Elizabeth R. DeSombre and Samuel Barkin, *The Turbot War: Canada, Spain, and Conflict over the North Atlantic Fishery*, Pew Case Study no. 226 (Washington, DC: Institute for the Study of Diplomacy, 2000).

comprehensively regulated by individual states within EEZs. High-seas fisheries would be the core focus, but straddling and highly migratory stocks would also be important to address. Exceptions to this rule might be made for specific fisheries that are currently managed bilaterally or multilaterally and are exploited by vessels that lack the range or other specific capabilities to access other international stocks, although it would be preferable if these exceptions were not made and their successful regulation were fully integrated into the global system. The Pacific salmon fishery, currently managed jointly by Canada and the United States, is an example, but it is a peculiar fishery for a number of reasons.[5] It may be the case that few other fisheries will fit into this category. Any fishery that is neither managed entirely within an EEZ by a state able to effectively exclude nonparticipant vessels from its waters nor included in the global system would end up bearing the brunt of excess fishing capacity that is squeezed out of the global regulatory system.

A Global Fisheries Organization and ITQs

A global fisheries organization would need to play at least three roles for this ITQ system to work. The first is a relatively straightforward record-keeping and coordination role. In order to keep track of who owns permits to what at any given point in time, a single centrally managed and authoritative register is useful. This register might also act as a single authoritative location for registering all trading of permits. A centralized approach would improve efficiency by avoiding duplication of record keeping and trading facilities. It would also provide a single large and therefore relatively liquid exchange for permits rather than several smaller and therefore less liquid exchanges. Liquidity is more of an issue in a global ITQ system than it is in single-species systems because the number of potential kinds of trades is exponentially larger. For example, if a fisher wants to sell ten tonnes of quota for North Atlantic turbot and buy five tonnes of quota for Patagonian toothfish, it would be much easier to do if everything were traded in one place, rather than the quota being sold

---

5. See, for example, J. Samuel Barkin, "The Pacific Salmon Dispute and Canada–US Environmental Relations," in *Bilateral Ecopolitics: Continuity and Change in Canadian–American Environmental Relations*, ed. Philippe Le Prestre and Peter Stoett, 197–210 (London: Ashgate, 2006).

in Halifax and bought in Hobart.[6] A single centralized exchange, if the global fisheries organization managed it well, would also help to prevent both miscommunication and intentional foul play[7] in the management of permit records and trading.

The organization might also play some role in oversight of regional quota setting. In the system envisioned here, RFMOs would continue to determine the size of the overall catch—we are not proposing a centralized system for the setting of TACs. RFMOs already exist and have mechanisms and staff in place for getting and processing scientific input. These systems on the whole work fairly well. When TACs are too large, it is generally not because of inadequacies in the scientific advice, but because the states within the RFMO commission decision-making process set them higher than scientific advice recommends. Taking apart this system and replacing it with a new and untried system responsible for creating hundreds of quotas is unnecessary. The large number of quotas required for the system as a whole to work also argues against centralizing the setting of quotas. No single scientific or management panel can deal directly with the high number of species and the variety of local conditions; separate research and analytical bodies (as currently exist within most RFMOs) can most usefully fulfill this function.

A centralization of quota setting would also have the effect of centralizing political disputes.[8] A dispute over a particular quota might undermine agreement over the range of quotas required. This disagreement can (and sometimes does) happen at the RFMO level, but it is a less serious problem in a single RFMO than it would be at the global level. There is a useful analogy here with global trade politics and the WTO. Centralizing information gathering and dispute settlement functions in the one organization generate real efficiencies. But centralizing political negotiations

6. Hobart, Australia, is where CCAMLR, which regulates much of the Patagonian toothfish fishery, is headquartered, and Halifax (actually Dartmouth, a suburb of Halifax), Canada, is where NAFO, which regulates much of the turbot fishery, is located.

7. For an example of foul play in international trading of environmental permits, see Joshua Chaffin, "Cyber-theft Halts EU Emissions Trading," *Financial Times*, 19 January 2011.

8. J. Samuel Barkin, "The Environment, Trade, and International Organizations," in *International Handbook of Environmental Politics*, ed. Peter Dauvergne, 334–347 (London: Edward Elgar, 2005).

in rounds of comprehensive bargaining can have (and is currently having) the effect of preventing any agreement on new rules. In trade negotiations, however, even with this problem, existing rules continue to work adequately in the absence of a new agreement. Because fishing quotas are often decided anew every year or two, globalized disputes that threatened the entire system of rules would have a much more pernicious effect.

In practice, though, what the quota-setting process looks like would likely shift under a global ITQ system. There is still scope for some elements of oversight of RFMOs in a global ITQ system managed by a global fisheries organization. This global body would monitor RFMO quotas to identify particular organizations that are systematically failing to set sustainable TACs, publicizing information about stock health over time that would indicate which RFMOs are succeeding in managing stocks well. More important, however, the market value of the ITQs within each RFMO will also send that signal because the value of well-managed fisheries will increase and the value of poorly managed fisheries will decline.

Should these mechanisms fail to work and dysfunctional RFMOs fail to reform themselves, the global fisheries organization would offer its own expertise and assistance. It might have a bureau dedicated to the study of and assistance with RFMO best practices.[9] Such a bureau can be relatively small on the assumption that RFMOs can be assisted sequentially, meaning that the bureau would need only enough staff to help one or two at a time. The global fisheries organization might be given the power to force change on particular RFMOs in extremis, although negotiating this power might turn out to be more politically complicated than its utility would warrant, especially because the price of quota shares may itself be a strong enough indicator to create change.

The other oversight a global fisheries organization would perform is to identify lacunae in RFMO coverage of global fisheries. A global ITQ system, as already noted, needs to be comprehensive to be fully effective precisely because of the balloon problem. But RFMOs are designed to be geographically and species specific, not comprehensive. It is their function

---

9. Some major fishing states already do this sort of thing. For example, the Canadian government has published a set of recommended RFMO best practices. See Fisheries and Oceans Canada, "Recommended Best Practices for Regional Fisheries Management Organizations: A Model RFMO," 2009, http://www.dfo-mpo .gc.ca/international/documents/poster-eng.htm.

to manage effectively within their remits; it is not their function to identify gaps between their remits and those of other RFMOs. Centralized oversight is needed to identify these gaps. The global fisheries organization would identify specific fisheries that have some international component that is not covered by existing RFMO regulation as they become apparent or as commercial fisheries develop for previously unsought fish. These species could then be assigned to existing RFMOs, or in some cases (for example, with a regional cluster of species) a new RFMO could be created. Although the resulting coverage would likely not be perfect, it would be better than without centralized oversight, and more comprehensive in this case means more effective. At minimum, central oversight may well be able to deal with major and glaring gaps in coverage more quickly than would otherwise be the case.

In addition to record keeping and oversight, a global fisheries organization would coordinate enforcement mechanisms. The most obvious element of enforcement is ensuring that vessels within the system abide by the rules. Someone must ensure that vessels flagged in member states do not overfish their quotas. Under a global ITQ system, what constitutes enforcement itself would change because what is being enforced would be somewhat different. States would presumably still be charged with ensuring that their nationals follow whatever rules exist (including requiring that fishers have the proper permits to land or sell the fish they catch). Enforcement of quotas currently is entirely the responsibility of member states, which are responsible for taking action against ships flying their flags or against their own nationals who are caught breaking the rules.[10]

A global fisheries organization would play a useful role in this process by acting as a clearinghouse for information about rule breakers across RFMOs and as a high-visibility outlet for information about member states that consistently fail to live up to their responsibilities in enforcing the rules on their own nationals. If the global organization is coordinating trading of ITQs, it might even exact meaningful penalties on states deemed not to be sufficiently enforcing by refusing to allow nationals of that state to purchase quotas. This policy would give individual fishers in countries with a history of poor fisheries enforcement an incentive to push for better governance because governance performance at

---

10. Elizabeth R. DeSombre, *Flagging Standards: Globalization and Environmental, Safety, and Labor Standards at Sea* (Cambridge, MA: MIT Press, 2006).

the national level would allow fishers within that country access both to global stocks and to global markets. Or the global organization might refuse to let individuals fish for other species or in other areas if they are found to have failed to live up to their obligations in one system or for one species.

The organization might also be of assistance in developing monitoring systems where there are positive returns to scale. For example, it might help to develop an effective global positioning tracking system, in which long-range vessels that fish international stocks would be required to carry a transponder.[11] Such a system would make it easier for individual RFMOs to require tracking and would make it possible to track large fishing vessels across RFMO boundaries, especially because under a truly global ITQ system individual vessels would be able hold quota allocations in multiple fishing regions.

A less obvious element of enforcement would be the creation of a system that punishes free riders, those states that choose not to join a global ITQ system. This issue is technically not one of enforcement because failing to take on an international legal obligation is in most cases completely legal. But involvement of all states that register industrial-scale fishing vessels is essential for fisheries management generally[12] and particularly for the successful functioning of a global ITQ system.

Moreover, international law has for the past several decades been moving in the direction of seeing the global commons as something to be managed collectively rather than as a free-for-all.[13] International oceans and fisheries law in particular has clearly and progressively been moving in the direction of this requirement. UNCLOS (1982), for instance, requires states whose fishers fish in a given area to "cooperate to establish subregional or regional fisheries organizations" to manage those species.[14] The associated Straddling Stocks Agreement (1995) goes further, mandating

11. Lee G. Anderson and Mark C. Holliday, eds., *The Design and Use of Limited Access Privilege Programs*, National Oceanic and Atmospheric Administration (NOAA) Technical Memorandum no. NMFS-F/SPO-86 (Washington, DC: NOAA, November 2007), 87–88.

12. DeSombre, "Fishing under Flags of Convenience."

13. Elizabeth R. DeSombre and J. Samuel Barkin, "Turtles and Trade: The WTO's Acceptance of Environmental Trade Restrictions," *Global Environmental Politics* 2 (1) (2002): 12–18.

14. UNCLOS (1982), Article 118.

that states with an interest in straddling or highly migratory fish stocks join existing fishery management organizations (or work to create them where none exists) and agree to implement these organizations' regulations whether they are members or not.[15]

International law thus increasingly supports—or even mandates—the idea that states should act in concert to ensure sustainability in the global commons. International trade law in particular has also developed precedents to allow states to interfere with trade for purposes of environmental protection if this interference does not unnecessarily benefit domestic over foreign goods.[16] Although none of these changes allows participant states to force nonparticipant states into a global ITQ system, it does allow them to take action to discourage nonparticipation and to reward entry into the system. In other words, participant states can create what economists call "clubs," which are mechanisms for enforcing rules by allowing access to something of economic value only to those actors who are members of the club and agree to its rules.[17] In the case of a global ITQ system for fisheries, these things of economic value are, first, access to fishing quota itself and, second, ports and markets.

Whatever other rules are decided about trading quotas, it seems obvious that states would have to be members of the RFMO from which a given fishing quota is issued. Because the bulk of authoritative enforcement in international law is carried out by states through domestic legal processes pertaining to their own nationals, these states would need to be legally obligated at the international level to enforce quota for their flagged vessels or fishers. That itself creates a form of club good because fishers can purchase quotas only if their states are members of the relevant RFMO. So if fishers want to be able to access quota they would put pressure on their flag states to join (or simply reflag in a state that has joined), much in the same way they would put pressure on their flag states to enforce the rules once they have joined. That process itself will create increased membership in RFMOs and diminish the problem of free riding they have faced in the current system.

An additional mechanism for discouraging free ridership on the global ITQ system is to prohibit nonparticipants from landing fish in member

15. Straddling Stocks Agreement (1995), Articles 8, 13, and 19–23.

16. DeSombre and Barkin, "Turtles and Trade."

17. J. M. Buchanan, "An Economic Theory of Clubs," *Economica* 32 (1965): 1–14.

countries' ports and selling fish to member countries' markets. There are precedents for both mechanisms in existing fisheries management systems that might be adapted to the global system. The most well-developed program is operated by ICCAT. Beginning in 1992, ICCAT required that bluefin tuna imported into member states be accompanied by a "statistical document" indicating where, when, and how it was caught, validated by an official of the vessel's flag state.[18] The program was expanded to include swordfish and other tuna species as well. More important, member states are required to exclude these fish without such a document or if the document does not demonstrate that they were caught within ICCAT regulations. This program has resulted in some new states joining the organization and other nonmembers ceasing their operations in the fishery.[19]

A similar program operates in CCAMLR, with a "catch documentation scheme" for Patagonian toothfish (also known as Chilean seabass) following rules and processes similar to the ICCAT approach. CCAMLR faces a few additional difficulties in implementing this program, however, because although it regulates catches in a circumscribed area, these fish also exist outside of that area. Where a fish has been caught, therefore, determines whether it is subject to CCAMLR rules. In order to address that issue, CCAMLR has created the Vessel Monitoring System, designed to allow tracking of fishing vessels to be able to determine where a fish was caught. As in the ICCAT program, CCAMLR's catch documentation process has led additional states to join the organization and others to cooperate with the commission even though they have not joined.[20]

This mechanism works in existing RFMOs only when it is implemented by states that are major markets for fish; likewise, on the global level, such an approach would work only if a critical mass of the world's states, in particular those that are the biggest importers of fish, participate. Persuading states to take on rules by disallowing access by their fishers to markets will work only if those states are the desirable markets for fish products. But a small number of countries account for a large majority of the world's fisheries imports. Just the EU, the United States, and

---

18. ICCAT, *Recommendation by ICCAT Concerning the Bluefin Tuna Statistical Document Program*, Recommendation 92-1 (Madrid: ICCAT, 1992).

19. DeSombre, *Flagging Standards*, 160–166.

20. Ibid., 169–175.

Japan account for more than two-thirds of global fisheries imports. Add in Canada, Norway, Australia, and Switzerland, and the figure gets close to three-quarters. With China and Russia, ten markets account for more than 83 percent of global fisheries imports.[21] Were all these countries to join a global ITQ system, fishing outside of that system would become much less lucrative for anyone who chose to engage in it. And without access to the ports of those countries, fishing in much of the world would become more expensive.[22]

As more states joined, the market for fish from nonparticipants would continue to shrink, thereby increasing incentives for the remaining free riders to join. In other words, getting a strong club going in the first place might be difficult, but, once begun, the club and its effects would likely grow stronger over time. And although such a system is designed to entice participation, it can (and the existing programs do) also serve the more traditional goal of enforcing compliance to the extent that it requires fishers to demonstrate their adherence to existing rules in order to land, transship, or sell their fish. The system would not entirely eliminate illegal fishing but would likely make it less profitable and therefore less common.

These approaches to enticing free riders to join the international regulatory system, and creating incentives for rule abiding, are not likely to create perfect implementation or compliance, but to expect a perfect enforcement mechanism for international rules, especially those pertaining to a global commons problem, is unrealistic. The better question is whether enforcement of ITQs under the auspices of a global fisheries organization would work better than the current pattern of adoption and enforcement of RFMO rules. And the answer, very likely, is yes. It will do so even if enforcement on a ship-by-ship basis does not improve. Embedding that enforcement in a universal system of tradable permits will improve implementation and effectiveness overall by filling in some of the cracks in the system and by mitigating the balloon problem. And this improvement would be in addition to the incentives an ITQ system creates for more effective management decisions.

---

21. As of 2007, the EU is counted as one market because it is, for purposes of international trade, legally one market. China includes Hong Kong. Figures are by landed value and are from FAO, *FAO Yearbook: Fishery and Aquaculture Statistics, 2007* (Rome: FAO, 2009), 42.

22. DeSombre, "Fishing under Flags of Convenience."

The creation of clubs also shows returns to scale: one big market exclusion club is more effective an incentive than a set of smaller, species-specific clubs because it prevents vessels from avoiding the club restrictions by changing species or regions.[23] In other words, enforcement on a ship-by-ship basis does not need to be better than existing RFMO enforcement for a global ITQ system to yield more sustainable fisheries governance. It simply needs to be no worse.

We do not address here a number of details of the specific functioning of a global fisheries organization in the management of an international ITQ system because they can be resolved in a number of ways without undermining the system's effectiveness. One of these details, as has already been noted, is the distribution of permits. The initial allocation of permits can be done in a variety of ways, and it would even be possible to leave to individual states in an RFMO the option to distribute their permit quotas as they see fit once initial regional quotas have been determined (although distribution of quotas at below-market rates to existing fishers would be suboptimal because it would constitute a subsidy and thus could not be repeated with future quotas). Alternately, national quotas might be eliminated and permits might be auctioned directly to fishers. This latter method would be more economically efficient but is probably less politically viable. What happens to permits after they are first distributed, however, is much more important to the effective functioning of a global ITQ system than the method of distribution in the first place.

Another set of details involves the specific market characteristics of the quota permits. Decisions would need to be made about the tenure of the permits. They need at minimum to be long term or renewable with a strong bias toward existing rights holders in order to have the desired effects on the incentive structure facing fishers (as discussed in chapter 9). Permanent permits have the advantage of being a stronger property right and therefore of having the stronger effect on individual incentives. Renewable permits have the advantage of making it easier to deny rights to chronic rule breakers (because nonrenewal is easier to effect legally than is the removal of a permanent permit). A nonpermanent system might also provide protection against the possibility, as in the European carbon

---

23. Richard Cornes and Todd Sandler, *The Theory of Externalities, Public Goods, and Club Goods* (Cambridge: Cambridge University Press, 1996).

market,[24] that initial market design might run into unforeseen problems. Both approaches, in other words, have advantages and disadvantages.[25]

The question of whether permits might be leased would similarly need to be addressed in negotiations. Leasing would allow fishers to essentially rent a quota for a year or two, without the capital investment of buying the quota. This option would increase flexibility for fishers and make permits more like normal property. But it would also make the system more complicated to oversee and would allow investors who are not fishers to buy permits as a leasing revenue source. Should states prefer that the permits be held by actual fishers rather than by financial investors, leasing becomes a problem. As is the case with the distribution and tenure of permits, an international ITQ system can work effectively whichever decision is made.[26]

And, of course, all the other specific rules about who holds a quota and the conditions under which it can be traded would have to be decided. Many of these rules, as discussed in chapter 9, pertain to any ITQ system, whether domestic or international, and there are many acceptable ways to make these determinations. But some decisions would have different implications at the international level. For instance, many ITQ systems have rules designed to minimize concentration of shares. The international level introduces a new element to that issue: Should there be rules to prevent concentration not just among any one type of permit holder or vessel, but across states? Given states' decision-making roles within RFMOs, too much concentration of shares in any one state would affect the political process of TAC determination. But the more rules limiting forms of trading, the less complete the property right, thus diminishing the system's incentive advantages. These structural aspects would need to be set up through the negotiation process that creates such a system (and

---

24. Frank J. Convery and Luke Redmond, "Market and Price Developments in the European Union Emissions Trading Scheme," *Review of Environmental Economics and Policy* 1 (1) (1997): 88–111.

25. Christopher J. Costello and Daniel Kaffine, "Natural Resource Use with Limited-Tenure Property Rights," *Journal of Environmental Economics and Management* 55 (1) (2008): 20–36.

26. Kate Bonzon, Karly McIlwain, C. Kent Strauss, and Tonya Van Leuvan, *Catch Share Design Manual: A Guide for Managers and Fishermen* (Washington, DC: Environmental Defense Fund, 2010), 61, http://www.edf.org/sites/default/files/catch-share-design-manual.pdf.

most likely amended over time in response to unforeseen effects), so we do not advocate specific solutions at this point.

A final set of logistics needing to be worked out involves the relationship between an international ITQ system and the management of national fisheries. In principle, there is no reason to involve fisheries in the international system if they can be effectively regulated by single countries. And it may well be the case that keeping national regulation distinct from the international system would make negotiations simpler by avoiding questions of national sovereignty and authority within EEZs. However, not all states have the administrative and monitoring capacity to regulate effectively within their own EEZs. Some species regulated by states (such as turbot) also exist in the high seas and are regulated there by RFMOs (such as NAFO);[27] some RFMOs, such as ICCAT, regulate species even within member states' EEZ.[28] In these contexts, it may benefit the states in those situations and the cause of fisheries sustainability overall to give states the ability to associate national regulatory systems with the global ITQ system.

This kind of domestic–international integration would likely be most appealing to states if leasing of permits were permitted; a state would otherwise lose the rent from its domestic fisheries as soon as permits were allocated in the first place. A state that leased the right to fish in its EEZ through the global fisheries organization would benefit from the ITQ system's enforcement mechanisms—interlopers would not be able to sell illegally caught fish on the international market. For example, states in West Africa that currently sell access rights to the EU but lack the capacity to monitor EU fishing within their EEZs might piggyback on a global ITQ system to ensure that EU vessels are not offloading more fish or different kinds of fish than they are supposed to be catching. Again, though, these details would ultimately be determined by international negotiation.

Is It Feasible?

The specific details discussed previously will need to be worked out in actual negotiations. Additional aspects of difference between any domestic

---

27. David R. Teece, "Global Overfishing and the Spanish–Canadian Turbot War: Can International Law Protect the High-Seas Environment?" *Journal of International Environmental Law and Policy* 89 (1997): 89–125.

28. ICCAT, 1966, Article 1.

and international system also need to be addressed in order to create a workable global ITQ system. One is the practical difference between domestic and international law. Domestic law is authoritative. Once a law comes into being, it can be imposed on all members of society; people do not have to consent directly in order to be bound and do not have to agree to enforcement measures to be subject to them. International law is not binding on those that have not voluntarily taken it on, and they need not only to agree to any substantive rules, but also to processes for enforcing those rules. As noted earlier, however, international law is changing in a way that makes it more accepting of comprehensive schemes to manage the global commons and makes it easier to encourage recalcitrant states to participate.

The use of market mechanisms that use trade restrictions in particular to encourage participation in such schemes has a history going back fully one-quarter of a century to the signing of the Montreal Protocol in 1987,[29] and the WTO Secretariat has remarked that existing schemes in ICCAT and CCAMLR to prohibit trade in regulated species with nonmember states "provide examples of appropriate and WTO-consistent (i.e. non-discriminatory) uses of trade measures in multilateral environmental agreements."[30] Port–state exclusion is also well accepted in international law; states are allowed to exercise any condition they choose for entry of ships to their ports.[31] The exclusion of nonparticipants from the markets and from participants' ports in a global ITQ scheme is thus viable in the context of contemporary international law, including international trade law.

But will such mechanisms of exclusion work? As is often the case in international environmental politics, they will work only if states representing the major markets in the international marine fish trade can be convinced to participate and if a subset of those states come to care enough about the system that they are willing to commit some real political capital to it. A system that does not include, for example, the United States or Japan will not succeed because each of these two markets individually

29. Duncan Brack, *International Trade and the Montreal Protocol* (London: Royal Institute of International Affairs, 1996).

30. WTO, *The Environmental Benefits of Removing Trade Restrictions and Distortions: The Fisheries Sector*, Note by the Secretariat, WT/CTE/W/167 (Geneva: WTO, 16 October 2000), 9.

31. DeSombre, *Flagging Standards*, 87–98.

is big enough to undermine the incentive structure that market exclusion is designed to create (or, conversely, they are big enough to ensure that others care about access to their markets if they do participate). Should a group that includes the biggest markets get some momentum going, the incentive structure will become self-reinforcing. In other words, starting an international ITQ system will be a real political challenge. But if it can be started and can gain traction among consumer nations, it is likely to get stronger and more effective over time.

One piece of evidence for the feasibility of an international ITQ system is the increasing call, both in the academic literature and even within RFMOs themselves, for the creation of one. One of the earliest calls came from G. T. Crothers and Lindie Nelson from the New Zealand Ministry of Fisheries, who in 2007 proposed a global framework for regulating high-seas fisheries via defined property rights given to states.[32] Rögnvaldur Hannesson similarly explored the question of what would be necessary for individual rights in high-seas fisheries, proposing the importance of a clublike structure of RFMO membership.[33] James Joseph, director of the IATTC for thirty years, argued after his retirement for the adoption of new options for managing capacity in tuna fisheries, including rights-based options such as ITQ systems.[34] And the IOTC has officially begun to attempt the creation of a long-term quota allocation that would have more in common with property-rights systems than is usually the case with RFMO quota processes.[35]

The complexity of an international ITQ system also creates potential difficulties. This complexity is precisely the reason that a global fisheries organization is necessary. The role of such an organization can usefully be conceptualized through the lens of Coasian market perfecting, as per the neoliberal institutionalist literature on international

32. G. T. Crothers and Lindie Nelson, "High Seas Fisheries Governance: A Framework for the Future?" *Marine Resource Economics* 21 (2007): 341–353.

33. Rögnvaldur Hannesson, "Rights-Based Fishing on the High Seas: Is It Possible?" *Marine Policy* 35 (2011): 667–674.

34. James Joseph, *Past Developments and Future Options for Managing Tuna Fishing Capacity, with Special Emphasis on Tuna Purse-Seine Fleets* (Rome: FAO, August 2005).

35. Jeremy Noye and Kame Mfodwo, "First Steps towards a Quota Allocation System in the Indian Ocean," *Marine Policy* 36 (2012): 882–894.

organizations.[36] A global fisheries organization would reduce transaction costs by creating a single trading site and platform for permits. It would improve information flows both by acting as a single centralized clearinghouse for information on TACs, quotas, and permits and by identifying and publicizing best practices for RFMOs. And it would improve property rights by clarifying a single set of rights attached to permit holding.

An international ITQ system cannot solve all problems of IUU fishing. It would be most effective at cutting down on unregulated fishing because it would contain mechanisms specifically designed to identify unregulated fisheries and bring them into the system (or exclude them from markets if they refuse) and thus to encourage unregulated fishers to work within the system. It will cut down somewhat on the problems of illegal and unreported fishing by making it more difficult to land and sell fish caught outside of the system. But the monitoring and enforcement mechanisms needed to cut down even further on IUU fishing at the level of the individual fishery will continue to be managed by RFMOs and flag states, and the process of incremental improvement to these mechanisms that has historically happened somewhat sporadically will need to continue, ideally at a more consistent pace.

Would a Global ITQ System Succeed?

The discussion to this point has focused on how the mechanics of an international ITQ system might work and whether such a system can even be set up in the context of international law and the norms of international organization. Such a system, overseen by a global fisheries organization, is viable—it can be made to work on its own terms. But what effect would it have on the problems that it would need to address—the balloon problem and the overcapacity in the global fisheries industry that drives it?

A global ITQ system would deal with the balloon problem directly by bringing all international fisheries within the system of regulation. Excess capacity would not be able to find unregulated fisheries to exploit when squeezed out of regulated ones because there would be none. It is for this

---

36. For the seminal discussion of this approach, see Robert Keohane, *After Hegemony: Cooperation and Discord in the World Political Economy* (Princeton, NJ: Princeton University Press, 1984).

reason that the system needs to be comprehensive. It also needs to work in concert with national regulators to prevent excess capacity from international fisheries being squeezed into national waters and fishing those waters unsustainably. The system thus also discourages overcapacity by making it clear to potential entrants into the industry that IUU fishing will be more difficult to do profitably than is the case in the current regionalized system of fisheries management.

The global ITQ system would also put downward pressure on capacity in the long term, in much the same way as domestic ITQ systems are designed to limit domestic fishing capacity. It increases barriers to entry while at the same time decreasing barriers to exit, and it undermines the fundamental reason for increasing capacity. Barriers to entry are increased (although only for new entrants if permits are given away rather than sold at the system's genesis) by adding the cost of buying permits to other costs of entry, such as acquiring a vessel.

An ITQ system also helps to eliminate barriers to exit. Under existing systems in which the right to fish is valuable if you use it but has no value to you whatsoever if you cease to use it, fishers need to remain in the industry in order to earn income from their fishing capacity. It is for this reason that the pressure for subsidies when quotas are decreased or fisheries are closed is so high. But under an ITQ system, permission to fish has monetary value and is tradable. It thus acts as a sort of retirement plan for fishers. Selling a quota in a well-managed fishery can give vessel owners exiting the industry either capital to engage in other livelihoods or the wherewithal to stop working. They therefore gain an incentive to leave earlier and to leave a healthy fishery rather than hold on in the fishery until the relevant stocks are fished out.[37] In other words, by creating a form of individual and tradable property rights in international fisheries, a global ITQ system would correct some of the pathologies that keep too many people in the industry and lead to pressure for governmental support to allow them to remain.

A major advantage of the actual operation of a global ITQ system is that it removes much of the incentive for overcapitalization, at both the individual level and the national level. In traditional fishery management approaches, fishers are able to access a greater proportion of available

---

37. Bonzon et al., *Catch Share Design Manual*, 46.

fish if they are more heavily capitalized—if they have bigger vessels, bigger engines, more range, or more advanced technology, and so on.[38] If RFMOs set quotas without setting national shares (as was done in early fishery management efforts), the incentive for increased investment is on the national level—states want their fishers to be able to access a greater proportion of the available fish. If there are gaps in international regulatory coverage, states want to ensure that their nationals will be able to beat out others in the race to catch those unregulated fish. And even when RFMOs pass TACs and distribute them nationally, the individual fishers within a given state can gain a greater part of the national share if they can catch the fish first. All of these situations lead to an incentive to create the biggest and fastest fishing vessels and to fish as quickly as possible.

But in a global ITQ system, anyone with a quota is guaranteed the right to catch that amount of fish. If the system covers all fish stocks, states no longer need to subsidize their fishers to give them a competitive advantage in catching fish. The same is true on the individual level: there is no need to race to get the fish before someone else does, so the fisher does not need to invest in bigger vessels or fish in unsafe conditions.[39] This lack of derby-style fishing is also beneficial for fishers' income. If fishers are racing to catch the quota before it is used up, everyone brings fish to market at roughly the same time, causing a glut of supply and forcing down prices. If, instead, fishers can take their time because their quota access is guaranteed, they can bring fish to market at a more staggered and lucrative time.[40]

An ITQ system also changes practicing fishers' time horizons. The incentive under CPR conditions to race for the fish is replaced by an incentive to manage for the long term. This incentive is even greater in a tradable-permit system than it is when fishers have nontradable individual rights to a fishery. For any long-term individual quota, the incentive is based on fishers' ability to continue working into the future and on the guarantee of personal access (unlike in a system in which quotas are not owned). In the case of a global ITQ system, the additional incentive is a substantial cash payoff at a discrete point in time (of the fisher's choosing)

---

38. Steven C. Hackett, *Environmental and Natural Resources Economics*, 2nd ed. (Armonk, NY: M. E. Sharpe, 2001), 93.

39. Ibid.

40. Bonzon et al., *Catch Share Design Manual*, 106.

when the fisher decides to sell. The economic value of the quota is dependent on the fishery's continued health. Tradability, in other words, gives fishers the ability to choose between continuing to fish and monetizing the future rent from a well-managed fishery. There is evidence to suggest that fishers can have higher discount rates for future fishing than investors have for financial assets.[41] Making fishers into investors as well by turning their fishing permits into financial assets thus should have the effect of increasing the degree to which they are concerned with the fishery's long-term health.

The increasing length of fishers' time horizons in this system would also help to undermine the effects of regulatory capture. The interests of fishers and the broader population would align more closely around sustainable management. A comprehensive ITQ system in the context of international agreement to eliminate subsidies would also decrease the expected returns of focusing lobbying efforts on extracting economic resources from general government. If the industry expected a lower return from lobbying and a higher return, via tradable permits, from sustainable management, it would focus less on the former and more on the latter. Finally, as the industry becomes smaller over time (as the resource mandates), and as the ITQ system makes it seem more like a regular industry and less like a romantic throwback, its ability to extract resources from general government sources would likely decrease over time in most countries. Maintaining and improving an international ITQ system and enforcing sustainable quotas within that system should therefore get easier over time once the system is established.

But What About . . . ?

Two particular sets of concerns might be raised against the creation of such a system. The first is whether it would be able to address some of the traditional problems that plague the current regulation of international (and domestic) fisheries, including the difficulty setting quotas and the management of things such as bycatch and high grading that can undermine the effectiveness of quotas. And the second involves the system's distributional effects.

---

41. Frank Asche, "Fishermen's Discount Rates in ITQ Systems," *Environmental and Resource Economics* 19 (2001): 403–410.

An international ITQ system, as noted earlier, would rely on existing RFMO mechanisms for the creation of TACs. The discussion of RFMOs in chapter 4 argues that these mechanisms are prone to generating TACs that are higher than those recommended by scientific panels. An international ITQ system would not fix these problems at the microscale. And, in fact, the new system would not be relied upon to save particular species that are being overfished because of a political failure of a particular quota-setting system. The system's goal exists at the macroscale—to relieve systemic pressures on global fisheries, not to save particular threatened stocks. The potential for deadlock and administrative overload that would be created by centralizing quota-setting outweighs, in our opinion, the gain of a centralized mechanism for setting TACs. An international ITQ system is imperfect in the sense that it does not directly address some of the microscale problems in global fisheries regulation, but it is not designed to do so.

At the same time, addressing the problems of macroscale regulation should over the medium and long terms take some of the political pressure off the process of quota setting. It would do so in part by denationalizing fishing quotas. If permits can be traded across nationalities and flags (within the set of nationalities and flags of member states), national representatives at RFMOs should become less concerned with maximizing national quotas. In an established fishery, in fact, the idea of a national quota may become entirely obsolete. And absent a concern with national quotas within a TAC, the incentive to push the TAC beyond what the scientists recommend should decrease. Furthermore, as fishers themselves develop longer time horizons in response to ITQs' incentive structures, national industries should be less focused on lobbying national fisheries regulators for larger TACs in international fisheries.

The same effect should be generated by the decreased incentive for capitalization and subsidization, as discussed earlier. Both of these phenomena drive the political pressure to continue high TAC levels even when the status of the fish stocks (and scientific advice) suggests otherwise. If ITQs decrease the pressure for overcapitalization and subsidies, these effects should in turn also decrease pressure for unsustainable TAC levels.

Then there are the questions of bycatch and high grading, which plague any fishery management system and are thus not unique to—or necessarily solved by—an ITQ system. There are reasons to suspect, however, that a global ITQ system would be no worse and may likely be better at addressing these problems than a traditional short-term nontradable quota system. As chapter 9 suggests, the diminished incentive to race for fish might lead to decreased bycatch because fishers can more carefully target the species they actually seek.[42] If an ITQ system requires permits for all landed fish, it increases the incentive to fish carefully and target only the desired species. The incentive to catch more than your quota and throw back less-desired species (known as "high grading") might also similarly decrease under an ITQ system; in practice, high grading has not affected most ITQ systems,[43] which may be because fishing more than your quota, only to discard some, takes time and money,[44] and in a system where you are guaranteed access to a share of the stock, doing so may cost you more than you gain in additional value. Monitoring systems are nevertheless important for any system to prevent bycatch and high grading.

A second set of concerns that can be raised about an international ITQ system as proposed here is that it might in practice favor the concentration of the international fishing industry in a way that harms small-scale and artisanal fishers as well as fishing industries in developing countries. There are a number of interrelated issues in this set of concerns that can be addressed separately, including two issues of scale—the scale individual vessels (artisanal and local-scale fishing) and the scale of firms (industrial concentration)—and two issues of distribution, across countries in general and between developed and developing countries.

The least problematic of these concerns from the perspective of the politics of creating an international ITQ system and a global fisheries organization is that of the scale of individual vessels. To a certain extent, the concern is misplaced. Most artisanal fisheries are near shore and therefore likely to be regulated nationally rather than internationally, whereas most international fisheries are far enough from shore that they can be

---

42. R. Arnason, *On Selectivity and Discarding in an ITQ Fishery: Discarding of Catch at Sea*, Working Paper no. 1 (Reykjavik: Department of Economics, University of Iceland, 1996); National Research Council, *Sharing the Fish: Toward a National Policy on Individual Fishing Quotas* (Washington, DC: National Academies Press, 1999).

43. National Research Council, *Sharing the Fish*.

44. M. C. Kingsley, "Food for Thought: ITQs and the Economics of High-Grading," *ICES Journal of Marine Science* 59 (2002): 649.

accessed only by industrial-scale vessels.[45] Furthermore, a key factor driving the building of larger fishing vessels is government subsidies. As the level of subsidization declines and ITQs insert increased market discipline into international fisheries, part of the incentive for increased vessel size will disappear.

Even beyond these factors, though, there are ways in which an international ITQ system favors artisanal over industrial fishing. In the current system, in most cases both industrial and artisanal fishers get access to stocks for free, giving an advantage to industrial approaches that take in greater quantities. With an ITQ system, however, there is a price for access, which increases the costs for industrial fishers more than for artisanal fishers. Furthermore, within any given species, a permit will be for a certain weight of fish, and therefore the system favors fishers who can command a greater price per weight. One of the advantages of many forms of artisanal fishing is that they deliver a higher quality of fish. Less damage is done to fish in the process of line fishing, for example, than trawl fishing.[46] And more artisanal fishing styles can also give fishers more discretion in what they catch. As such, artisanal fishers pay the same amount per fish to purchase fishing quota as operators of larger vessels using more industrial fishing techniques, but they can often charge a premium on their catches in the market. They may therefore have a cost advantage on top of any savings generated by lower capital requirements for equipment. This advantage can be increased by specific incentives offered either by individual governments or, by agreement, by all members of the system for the use of artisanal techniques.

Furthermore, to the extent that a preference for smaller-scale fishing is driven by a preference for the smaller environmental footprint per amount of fish caught by smaller vessels, those techniques can be encouraged or mandated by direct regulation rather than through the ITQ system. RFMOs or other international cooperative mechanisms (including a global fisheries organization) can simply prohibit environmentally damaging fishing techniques, as has already been done quite successfully, for example, with driftnets. It is worth reiterating in this context that an

---

45. For a comparison of industrial and artisanal fishing, see Elizabeth R. DeSombre and J. Samuel Barkin, *Fish* (Cambridge: Polity Press, 2011), chap. 3.

46. Paul Greenberg, *Four Fish: The Future of the Last Wild Food* (New York: Penguin, 2010).

international tradable ITQ system augments rather than replaces existing regulation of global fisheries. To the extent that a comprehensive ITQ system ends up generating a higher level of state participation than current RFMO management, the result will be a higher proportion of global fishing effort happening within these regulatory structures and less unregulated fishing.

Beyond the scale of individual vessels, the scale of firms within the fishing industry might also be a concern. All else equal, ITQ systems are prone to industrial concentration. Although it is true that all trading is voluntary, in practice individuals and small firms may be more likely to have short-term economic needs that give them reason to sell quota, and big firms often have easier access to the capital needed to buy permits. In the Icelandic system, for example, the share of quotas owned by the largest companies doubled in the system's first ten years;[47] further concentration is likely prevented primarily by a fear of government regulation should further concentration occur. From a pure economic perspective, concentration is not a problem until the point at which it allows oligopoly/oligopsony power by the remaining firms in the industry. Short of that level of concentration, bigger firms can be expected to be more efficient by rationalizing nonfishing functions such as maintenance and sales and spreading risk across more vessels.[48]

But the concern regarding concentration is usually not about economic issues (or ecological ones), but about social issues. Should governments decide that maintaining existing fishing communities is a priority and that the concentration of permits in the hands of a few big firms should be minimized, there are strategies they can use. These strategies may come at the expense of economic efficiency, but they should not come at the expense of the ITQ system's sustainability effects. The most straightforward of these strategies is to allow only individual proprietors of fishing vessels to buy permits,[49] which can be done either at the national level or, by agreement among member states, at the international level. If the latter, it can be done with respect to specific fisheries or types of fisheries, or it can be written into the ITQ system's basic rules. Certain restrictions,

---

47. Einar Eythórsson, "A Decade of ITQ-Management in Icelandic Fisheries: Consolidation without Consensus," *Marine Policy* 24 (2000): 483–492.

48. Bonzon et al., *Catch Share Design Manual*, 46.

49. Ibid., 107.

short of banning corporate ownership altogether, can be used that limit the number of permits any one corporate entity would be able to own.[50] The details of such restrictions should not affect the system's ability to limit overcapacity and the balloon problem.

One problem that might arise if rules to limit industrial concentration were employed by one state or by a subset of states within the system is that this limitation would generate incentives for corporate entities interested in acquiring permits to register ships in other states. This problem is part of the question of the distribution of fishing permits across states, which is something that will ultimately be determined as the system is created. From a purely economic perspective, the idea of quota shopping across states is not a problem. In fact, it is part of the conceptual framework of a tradable-permit system: fishers will buy permits wherever the industry can be run most efficiently. From a sociopolitical perspective, it may be more problematic, however; some states may be unwilling to see a large proportion of their fishing industry migrate elsewhere. This phenomenon is already a problem under current forms of global management; more than 20 percent of open ocean fishing vessels currently are registered under FOCs,[51] many of which operate outside the international regulatory system altogether. Although the problem is not unique to an ITQ system, it might be able to be addressed by the design of the system either if rules about concentration were set globally or if rules mandated that certain percentages of quota be retained within specific states.

It should also be noted that states would retain the ability to restrict domestic fishing to local vessels—the ITQ system would apply only to international fisheries. And states would presumably retain the right to require that vessels fishing for straddling and migratory species within their EEZs be registered locally, even if those species are covered by the global system. States would still therefore be able to require local registration of fishers for fisheries representing a significant majority of the world's catch.

A final concern is that a global ITQ system will favor developed over developing countries. To a certain extent, this concern is misplaced. Given the national governments' continued control of fishing within EEZs, open

50. Ibid., 106–107.
51. DeSombre, "Fishing under Flags of Convenience."

trading in permits primarily affects long-range deep-sea fishing, an undertaking that is in any case beyond the poorest countries' means. And the advantage of lower labor costs in middle-income countries may well outweigh any advantage of capital availability held by fishers in developed states.[52]

It is also the case that an ITQ system offers access opportunities for fishers in developing states that are greater than exist under the current system, in which national quotas, often allocated by fishing history, make it difficult for new members in RFMOs to gain access to catches. Under a system in which a quota can be sold, it can be bought. And funding for fishing quotas might even be part of an economic development scheme by development organizations or by international funding entities such as the GEF. Developing countries may therefore be less disadvantaged under a global ITQ system than they would be under the current RFMO system. And it would be less problematic for the system, too, and therefore better for the developing states themselves.

But even beyond these factors, the concern is based to a significant degree on a false assumption that the development of fishing industries is a useful means of economic development. As argued in chapter 7, the use of fisheries development as a form of economic development causes more problems than it solves. To the extent that an international ITQ system prevents governments from putting resources into a long-range fishing fleet that will never be profitable, it does the taxpayers of those countries a service.

Nonetheless, the North–South politics of the issue may prove problematic, but compromise might be possible. In exchange for agreeing to the system, developing countries might be promised increased assistance in the development of their domestic fisheries' regulatory structure. This assistance could be provided by existing international organizations, such as the World Bank through its existing programs for improving fisheries governance, and could be financed with savings from decreases in fisheries subsidies globally. In this way, concerns by developing countries about being frozen out of an international ITQ system could be leveraged into improved fisheries management in developing-country EEZs as well as in

---

52. For a nuanced discussion of issues of cost structure and competitiveness in international fisheries, see D. G. Webster, *Adaptive Governance: The Dynamics of Atlantic Fisheries Management* (Cambridge, MA: MIT Press, 2009).

international fisheries. And international access via development funding would be much less problematic under an ITQ system.

Conclusions

Although the negotiation process that creates a global fisheries organization would need to work out the details of how a global ITQ system overseen by a global fisheries organization would operate, the advantages of regulating global fisheries in this manner are clear. Long-term tradable quotas give fishers a long-term investment in the sustainable management of global fisheries. Tradable permits increase new actors' ability to enter the fishery and remove the barriers to exit for those who otherwise might continue to fish beyond what the stock can support. An ITQ system decreases the incentive for overcapitalization and subsidies and therefore decreases the likelihood of regulatory capture that accompany those characteristics. The actual experience of domestic ITQ systems is promising: they have shown great ecological benefits[53] and have led to economic benefits both nationally and individually.[54]

Although some concerns may be expressed about disadvantages to developing countries or small-scale fishers, there is no reason to assume that the system we propose would be any worse for these entities than the current regulatory system, and an ITQ system may on its own be more advantageous for them than the status quo. The design of the ITQ system and its interaction with international development organizations would provide additional opportunities to assist economically disadvantaged populations.

---

53. Christopher Costello, Steven D. Gaines, and John Lynham, "Can Catch Shares Prevent Fisheries Collapse?" *Science* 321 (2008): 1678–1681.

54. Ragnar Arnason, Kieran Kelleher, and Rolf Willmann, *The Sunken Billions: The Economic Justification for Fisheries Reform* (Washington, DC: World Bank, 2008).

11

Conclusion

The first step in saving global fisheries is to acknowledge that the current approach to regulating marine fishing is not working. Fisheries are increasingly depleted, and fishing industries need to be highly subsidized to support fishers in an increasingly unproductive endeavor as existing numbers of fishing vessels chase fewer and fewer fish. Although some RFMOs have had some success protecting particular fish stocks, they have done so in a context of constant or growing overall fishing capacity, and successful protection in one area simply shifts fishing effort to other areas or species.

We argue for the creation of a new global fisheries organization as a mechanism to reorient the way in which global fisheries are managed. Many details remain to be worked out. A new organization, even with an effective agreement on eliminating subsidies and a working international ITQ system, would not be a perfect solution to the problem of global overfishing. But it would be more effective in combination with the current system of microscale regulation based on RFMOs than the latter system alone because it would address the central macroscale problem that cannot be addressed from within the current system: chronic overcapacity in the fishing industry.

Any reduction in excess capacity makes microscale management more effective, and the more excess capacity is reduced, the more effective existing regulatory systems can be. Less fishing capacity takes some of the air out of the balloon problem by limiting the extent to which fishers forced out of one fishery compete for access to others. And it reduces the incentive to fish outside of the regulatory system by balancing resources available in the system with the capacity that is chasing those resources. In other words, it allows everyone in a smaller industry to make a living within the system. As such, even if enforcement levels at the microscale do not improve, adding macroscale regulation will make these levels more effective by reducing the incentives and the pressures to cheat.

Reducing overcapacity may at first seem a fairly straightforward thing to do, but creating an effective system for doing so in the setting of international politics is not simple. It will require getting the economics right by addressing the ways in which fish are different from most economic goods. It will require getting the politics right by creating regulatory structures and systems that are less easily undermined by regulatory capture than those currently in place and that reduce the likelihood of regulatory capture over time. It will also require changing the way we think about fishing as an industry, without which even the right economics and the right politics will over time be undermined by social claims that create a vicious political cycle in which subsidizing the fishing industry increases its political voice and its continued demand for support at the expense of the fisheries' health.

The Economics

Environmental economics can be a useful tool in resource management. It provides techniques for aligning the incentives facing individual resource extractors with social needs for sustainable stewardship. But fisheries regulation at both the domestic level and the international level has until recently failed to make use of these techniques. Several countries, however, have begun to make use of mechanisms to better align regulation with fishers' economic incentives—from vessel buybacks to the assignment of property rights over fish resources to individual, corporate, and community actors.

These experiments have had mixed results. Buybacks, for example, can be problematic when fishers who sell their vessels do not lose their rights to fish because they may instead take up fishing from another boat or in another location. And when the boats bought back are not scrapped, they may end up in another fishery, adding to capacity elsewhere. Individuals have an incentive to sell their vessels, but less incentive to leave the industry and take their capital with them. Aligning incentives, although conceptually straightforward, can be complicated in practice.

One of the techniques suggested most frequently in the environmental economics literature for aligning individual incentives with the social good is the assigning of clear property rights, often referred to as "privatization." But privatization is a complicated strategy in international politics, with its two levels of actors—states and individual people. The creation of EEZs privatized from a state perspective, putting under state control waters that had previously been part of the global commons. But because most states re-created open-access approaches within these state-run waters, albeit for only their nationals or others who paid for fishing access, they re-created a tragedy of the commons on a smaller scale.

Some of the more recent national efforts at fisheries regulation that have focused on property rights have been quite promising, in particular mechanisms such as ITQs. These approaches are designed to align individual fishers' incentives with the demands of sustainable stewardship of the resource by giving the fishers an individual stake in the fishery's long-term health. These efforts can serve as a useful model for an international system to the extent that they can be adjusted to the realities of international politics.

There is, however, a limit to the extent that ITQs or other systems designed to mimic privatization can fully align private incentives with the public good. To the extent that the public good is defined as sustainable management of the world's fisheries, it is an environmental rather than an economic good. Privatization therefore aligns incentives perfectly only if fishers do not discount the future. The greater the discount rate, the less well aligned are the incentives. Management systems that mimic privatization need therefore to be supplemented by more traditional forms of command-and-control regulation as well as by effective monitoring and enforcement. A system designed to have the effect of privatization on international fisheries would therefore need to complement rather than replace existing RFMO-based regulation.

The Politics

Even a regulatory system that gets the environmental economics right and that finds the right mix of market incentives and command-and-control rules and the right mix of macro- and microscale management will ultimately see its effectiveness undermined if it fails to get the politics right.

Market incentives can work only if the actors at whom the incentives are targeted are sensitive to market signals. Regulatory capture has been used to protect actors from the market. Provision of subsidies, for instance, allows fishers to continue to fish even when the economic value of the fish they catch does not otherwise justify their effort. Regulators' focus on maximizing industry revenue in the short and medium term perpetuates the industry's political voice at the cost of sustainable management.

Attempts to reduce the size and capacity of the industry are stymied by regulatory capture, which is common in the domestic fishing sector and is a problem in national participation in international regulatory regimes. The fishing industry prefers to maintain its size and income base. Its ability to capture rents, in forms ranging from subsidies to unsustainably high quotas that transfer value from future populations to current fishers, has the ability to dilute or even cancel any market signals generated by property-rights regimes based on environmental economic analysis. Any system to oversee global fisheries regulation must be designed to overcome, co-opt, or avoid opposition from national regulators and from fishing industry actors that currently exercise various levels of influence over those regulators.

An ITQ system has the potential benefit of buying the industry into a new regime with less scope over the long run for regulatory capture that undermines effective management. Many existing ITQ systems do this by giving away quotas to existing fishers rather than making them buy the quotas, thus offering incumbents in the industry a major advantage over outsiders. This approach is both economically and fiscally inefficient; the literature in environmental economics on tradable permits generally argues in favor of auctioning permits rather than giving them away to industry incumbents.[1] But the "give away" approach is politically efficient. It gives the industry an incentive to accede to a new system, and once the system is in place, industry participants will develop vested interests in it. Furthermore, once the system is in place, it changes the incentives of a captured regulator from subsidizing and overfishing to protecting the value of the permits, which is accomplished by successful resource

---

1. Peter Cramton and Suzi Kerr, "Tradeable Carbon Permit Auctions: How and Why to Auction Not Grandfather," *Energy Policy* 30 (4) (2002): 333–345; Lawrence H. Goulder, Ian W. H. Parry, Roberton C. Williams III, and Dallas Burtraw, "The Cost-Effectiveness of Alternative Instruments for Environmental Protection in a Second-Best Setting," *Journal of Public Economics* 7 (3) (1999): 329–360.

management. Such a system also decreases the industry's political constituency by decreasing its size.

The Discourse

Even getting both the economics and the politics right will prove insufficient to generate sustainable management of global fisheries unless we change some of the ways we think about fishing as an industry. Fishing is often considered a way of life that needs to be protected and a good route to economic development. Both of these perspectives reinforce regulatory capture by increasing the claims by current and potential members of the industry on both marine and fiscal resources. And both undermine efforts at sustainable management by using social claims on the resource base to distract attention from the fact that the inherent material limits on the resource base cannot be changed by social claims. Fishing needs to be reframed as an extractive industry that can cause widespread environmental damage and as a counterproductive long-term strategy for economic development.

Fishing has a mystique in many countries, in particular developed ones, that is not shared by other industries, even extractive ones. This mystique is based on a vision of fishing that is both historically and economically inaccurate. Most fishing communities before the era of steel vessels and fossil fuels were marked by poverty, and many were notable for extraordinarily high occupational death rates.[2] What looks quaint in paintings was rather less appealing in practice. Fishers who participate in contemporary industrial fisheries are not remotely interested in returning to the danger and poverty of traditional distance fishing. But earning an income compatible with the standards of the contemporary economy is simply not compatible with maintaining traditional fishing communities intact. Should governments wish to keep their maritime communities intact, they should enable transitions out of industrial fishing rather than engage in ultimately futile (and expensive) attempts to keep an economically doomed industry operating.

---

2. For an evocative description of this phenomenon in the context of the Gloucester, Massachusetts, fishery, see Mark Kurlansky, *The Last Fish Tale: The Fate of the Atlantic and Survival in Gloucester, America's Oldest Port and Most Original Town* (New York: Ballantine Books, 2008).

Developing countries similarly need to recognize that the fishing industry is for the most part a development dead end. To them, it may seem at first to be an appealing opportunity. Fish, like other natural resources, are free for the taking, and the creation of a fishing industry may therefore seem a useful path to economic growth. But fish are not like other natural resources—the limits of extraction have for the most part already been reached, and therefore, with a few local exceptions, the industry is in the medium term a no-growth enterprise. It is small and so will not yield the sort of rents that can be used to subsidize development elsewhere. In fact, it is more likely to be a fiscal drain than a fiscal boon. Nor is it a viable source of employment growth in the medium term, given the inherent limits of the resource.

Calls for special treatment of developing countries in international environmental politics are understandable; these countries do not contribute to global environmental degradation to the same extent as rich countries and do not have the resources to adapt as quickly to environmental degradation or to a new regulatory environment. But the logic is different on this issue: the use of fisheries as a tool of economic development is not only bad environmental policy, but bad development policy as well. Developing countries should not seek to increase their fisheries development because doing so will not help their economies, especially in the context of current international approaches to fisheries management.

Changing ways of thinking on such issues takes time and perseverance. One way to do so is simply to participate in the discourse and to stress new ways of thinking. A key goal of this book is to participate in the process of reframing the debate. Beyond the scholarly discourse, however, changing the discourse in the relevant regulatory institutions can help. Signs of such a change can already be seen. The international development community, led by the World Bank, has increasingly come to recognize that subsidizing the development of fishing industries in poor countries is a bad idea. And some governments are coming to recognize that the only way to ensure a healthy fishing industry in the medium term is to limit its size. But the process of change can be accelerated with participation by institutional actors central to international fisheries governance.

A Global Fisheries Organization

One way to address the economics, politics, and discourse of sustainable global fisheries management all at the same time is to create a global fisheries organization. The creation of such an organization or even discussions toward the creation of such an organization would have a direct impact on the discourse of international fisheries regulation. It would constitute a direct recognition by participant countries that macroscale regulation of international fisheries is necessary and would by definition put the issue on the international political agenda. Bringing international organizations such as UNEP and the World Bank into the discussions would emphasize sustainable management, and bringing in funding bodies such as the GEF would help to smooth the path to participation for poorer countries that are cautious about losing a potential path to economic development. The creation of a global fisheries organization would have the effect of reinforcing this new approach to conceptualizing fisheries.

A global fisheries organization would not be immune to regulatory capture. Negotiations to create such an organization would succeed only if they worked in some way with, rather ignoring, current levels of industry influence on regulators. But once created, it would undermine regulatory capture by reducing over time the industry's size and political clout; in the context of management by the creation of property rights, it would also work to better align the interests of industry with the long-term sustainability of global fisheries.

To succeed, however, the organization would also have to get both the environmental and institutional economics right. It could address some of the inefficiencies in current regulation by centralizing negotiations so that all participants would agree where negotiations on global fisheries issues should be held and could centralize and rationalize information gathering and dissemination. It could act as an administrative and informational center for global fisheries regulation.

In order to have a major impact, however, such an organization would need to successfully address the problem of overcapacity. The primary ways it might address this problem would be via cooperation in reducing subsidies to the fishing industry and the creation of a macroscale quota system to control the balloon problem and provide incentives to fishers to participate in the reduction of capacity over time. These strategies might be undertaken sequentially, with subsidy reduction as a first step, followed by international ITQs, or they might be undertaken concurrently. The implementation of subsidy reduction might be facilitated by a new global fisheries organization, but it would be potentially more successful if negotiated through existing international organizations such as the WTO or even as a stand-alone agreement as long is it has the major fishing states' participation. An international ITQ system, conversely, would likely require a new institutional structure or a major modification of an existing structure such as UNEP.

In this book, we have sketched the outlines of a possible institutional structure for a global fisheries organization. These ideas are intended as suggestions rather than as a fully developed proposal. Some sort of new organizational structure is needed, but macroscale regulation of global fisheries is compatible with a variety of institutional forms, from an entirely new organization to modification of existing ones. In other words, we are not sketching out what we feel should be the institutional results of negotiations toward a new model of global fisheries governance. Rather, we are making the case that these negotiations need to happen and need to focus on macroscale regulation to complement existing patterns of and structures for microregulation.

It should be stressed again at this point, as has been noted elsewhere in this book, that we are claiming neither that our proposal needs to be adopted intact nor that it should be adopted to the exclusion of other ways of better managing global fisheries. An ITQ system without reductions in subsidies is problematic because the use of subsidies to buy ITQs would undermine incentives to reduce capacity and would result in competition by national regulators to outsubsidize other countries in purchasing them. But reducing subsidies would help even in the absence of the creation of tradable quotas. And changing the way we think about fishing as an extractive industry would reduce the pressure to subsidize, even in the absence of any formal agreement. At the same time, all of these changes would work in tandem with area-based management efforts such as the creation of marine protected areas and exclusion zones. And sustainable fisheries movements, which aim to educate consumers of fisheries products regarding the dangers of unsustainable exploitation of the oceans, will in all likelihood be a necessary component of any attempt to generate the political will to create mechanisms for macroscale management. All this being said, none of these approaches in isolation is likely to effectively address the problem of overcapacity in the global fishing industry. And management efforts are unlikely to be fully effective if they fail to address the capacity issue.

Would any of our proposals work? Are a global fisheries organization and a system of macroscale regulation within it politically feasible? Neither will be easy. Negotiations would fall prey to specific national and industry interests, and the protection of fish and ocean ecosystems is rarely at the top of national political agendas. But if the politics are done right, creating such a global system is not impossible. It would likely require some political compromise to move from the current regulatory context to the one we propose. For instance, fishers currently exploiting international stocks might be more amenable to a new property-rights regime if they received access to the first round of permits at less than market rates (or even for free), even though there are good economic and policy reasons to distribute permits by auction. Likewise, in lieu of differentiated obligations for developing countries, development organizations or the GEF might fund the purchase of existing ITQs for developing states; that approach would have the effect of increasing access by these states to fishing opportunities without increasing capacity. As well as these carrots, sticks in the form of port and market exclusion of nonparticipants might be used. And the political process might prove easier if existing international organizations involved in fisheries issues, ranging from RFMOs to broader environmental organizations to finance and trade organizations, were included in some way. Success would even then of course not be guaranteed. But we will certainly fail if we do not try.

A Call to Action

This is our core argument: we should try. Global fisheries and with them the health of the global marine ecosystem are approaching the point of crisis. Current regulatory efforts are not working, so we need to try something different. That something needs to address the problem of overcapacity in the fishing industry and needs to be global in scale in order to prevent the excess capacity from simply moving somewhere else. This book is designed to elaborate the problems with existing approaches to addressing fisheries issues and to start a conversation about where to go from here. It is easy to dismiss this call to action as politically naive or infeasible, but a macroscale regulatory system could be set up in a way that addresses the short-term interests of key current participants in international fisheries politics. It would therefore be something that those participants might take seriously if it were to develop political momentum. And the only way to generate that momentum is to take the idea seriously in the first place.

Fisheries may not be able to serve as a source of unimpeded development opportunities or an inexhaustible food source for a growing world population. But they provide important ecological functions, have economic and social value, and, if managed properly, can contribute to long-term sustenance and employment for many people. For them to do so, however, we must create global institutional structures that can work over the long term to save global fisheries.

Bibliography

Acheson, James M. *Capturing the Commons: Devising Institutions to Manage the Maine Lobster Industry*. Lebanon, NH: University Press of New England, 2003.

Acheson, James M. *The Lobster Gangs of Maine*. Lebanon, NH: University Press of New England, 1988.

Akiwumi, P., and T. Melvasalo. "UNEP's Regional Seas Programme: Approach, Experience, and Future Plans." *Marine Policy* 22 (3) (1998): 229–234.

Alder, Jackie, Helen Fox, and Miguel Jorges. "Overseas Development Assistance to Fisheries as a Subsidy." In *Catching More Bait: A Bottom-Up Re-Estimation of Global Fisheries Subsidies*, ed. Ussif Rashid Sumaila and Daniel Pauly, 54–67, Fisheries Centre Research Report, vol. 14, no. 6. Vancouver: Fisheries Centre, University of British Columbia, 2006.

Allen, G. C. *Japan's Economic Recovery*. New York: Oxford University Press, 1958.

Allen, Robin, James Joseph, and Dale Squires. *Conservation and Management of Transnational Tuna Fisheries*. Malden, MA: Wiley-Blackwell, 2010.

Alps, Robert, Laurence T. Kell, Hans Lassen, and Innar Liiv. "Negotiation Framework for Baltic Fisheries Management: Striking the Balance of Interest." *ICES Journal of Marine Science* 64 (2007): 858–861.

Alverson, David L., Mark H. Freeberg, Steven A. Murawski, and J. G. Pope. *Global Assessment of Fisheries Bycatch and Discards*. Food and Agricultural Organization (FAO) Fisheries Technical Paper no. 339. Rome: FAO, 1994.

American Sportfishing Association. "Government Affairs." 2009. http://www.asafishing.org/government/sustainable_fisheries.html.

Anderson, Lee G., and Mark C. Holliday, eds. *The Design and Use of Limited Access Privilege Programs*. National Oceanic and Atmospheric Administration (NOAA) Technical Memorandum no. NMFS-F/SPO-86. Washington, DC: NOAA, November 2007.

Annala, John H. "New Zealand's ITQ System: Have the First Eight Years Been a Success or a Failure?" *Reviews in Fish Biology and Fisheries* 6 (1996): 43–62.

Annala, John H., Kevin J. Sullivan, and Arthur J. Hore. "Management of Multispecies Fisheries in New Zealand by Individual Transferable Quotas." *ICES Marine Science Symposia* 193 (1991): 321–329.

Anticamara, J. A., R. Watson, A. Gelchu, and D. Pauly. "Global Fishing Effort (1950–2010): Trends, Gaps, and Implications." *Fisheries Research* 107 (2011): 131–136.

Aranda, Martin, Paul de Bruyn, and Hilario Murua. *A Report Review of the Tuna RFMOs: CCSBT, IATTC, IOTC, ICCAT, and WCPFC.* EU FP7 Project no. 212188 TXOTX. 2010. http://www.txotx.net/docums/d22.pdf.

Arnason, Ragnar. "Fisheries Subsidies, Overcapitalisation, and Economic Losses." In *Overcapacity, Overcapitalisation, and Subsidies in European Fisheries: Proceedings of the First Workshop of the EU Concerted Action on Economics and the Common Fisheries Policy (28–30 October 1998)*, 27–46. Portsmouth: University of Portsmouth, 1999.

Arnason, Ragnar. "On Catch Discarding in Fisheries." *Marine Resource Economics* 9 (3) (1994): 189–207.

Arnason, Ragnar. *On Selectivity and Discarding in an ITQ Fishery: Discarding of Catch at Sea.* Working Paper no. 1. Reykjavik: Department of Economics, University of Iceland, 1996.

Arnason, Ragnar. *A Review of International Experiences with ITQs.* Annex to *Future Options for UK Fish Quota Management*, Centre for the Economics and Management of Aquatic Resources (CEMARE) Report no. 58. Portsmouth, UK: University of Portsmouth, June 2002.

Arnason, Ragnar, Kieran Kelleher, and Rolf Willmann. *The Sunken Billions: The Economic Justification for Fisheries Reform.* Washington, DC: World Bank, 2009.

Arrow, Kenneth J. *Social Choice and Individual Values.* New York: Wiley, 1951.

Asche, Frank. "Fishermen's Discount Rates in ITQ Systems." *Environmental and Resource Economics* 19 (2001): 403–410.

Asia Pacific Economic Cooperation. *Study into the Nature and Extent of Subsidies in the Fisheries Sector of APEC Members' Economies.* Singapore: Asia Pacific Economic Cooperation, 2000.

Axelrod, Robert. *The Evolution of Cooperation.* New York: Basic Books, 1984.

Azar, Samia Atoine. "Measuring the US Social Discount Rate." *Applied Financial Economics Letters* 3 (2007): 63–66.

Barber, Willard. "Maximum Sustainable Yield Lives On." *North American Journal of Fisheries Management* 8 (2) (1988): 153–157.

Barclay, Kate, and Sun-Hui Koh. "Neo-liberal Reforms in Japan's Tuna Fisheries? A History of Government–Business Relations in a Food-Producing Sector." *Japan Forum* 20 (2) (2008): 139–170.

Barkin, J. Samuel. "Discounting the Discount Rate: Ecocentrism and Environmental Economics." *Global Environmental Politics* 6 (4) (2006): 56–72.

Barkin, J. Samuel. "The Environment, Trade, and International Organizations." In *International Handbook of Environmental Politics*, ed. Peter Dauvergne, 334-347. London: Edward Elgar, 2005.

Barkin, J. Samuel. "The Pacific Salmon Dispute and Canada–US Environmental Relations." In *Bilateral Ecopolitics: Continuity and Change in Canadian–Ameri-*

can Environmental Relations, ed. Philippe Le Prestre and Peter Stoett, 197–210. London: Ashgate, 2006.

Barkin, J. Samuel. "Time Horizons and Multilateral Enforcement in International Cooperation." *International Studies Quarterly* 48 (2) (2004): 363–382.

Barkin, J. Samuel, and Elizabeth R. DeSombre. "Unilateralism and Multilateralism in International Fisheries Management." *Global Governance* 6 (3) (2000): 339–360.

Barkin, J. Samuel, and Kashif Mansori. "Backwards Boycotts: Demand Management and Fishery Conservation." *Global Environmental Politics* 1 (2) (2001): 30–41.

Barkin, J. Samuel, and George E. Shambaugh, eds. *Anarchy and the Environment: The International Relations of Common Pool Resources*. Albany: State University of New York Press, 1999.

Barkin, J. Samuel, and George E. Shambaugh. "Hypotheses on the International Politics of Common Pool Resources." In *Anarchy and the Environment:: The International Relations of Common Pool Resources*, ed. J. Samuel Barkin and George E. Shambaugh, 1–25. Albany: State University of New York Press, 1999.

Barrett, Scott. *Environment and Statecraft: The Strategy of Environmental Treaty-Making*. Oxford: Oxford University Press, 2003.

Basch, Michael, Julio Pena, and Hugo Dufey. "Economies of Scale and Stock Dependence in Pelagic Harvesting: The Case of Northern Chile." *Cuadernos de Economia* (Santiago, Chile) 36 (1999): 841–873.

Bast, Joseph. "Welcome to the Heartland Institute." 2010. http://heartland.org/about.

Becker, Gary S. "A Theory of Competition among Pressure Groups for Political Influence." *Quarterly Journal of Economics* 43 (1) (2011): 45–65.

Beddington, J. R., D. J. Agnew, and C. W. Clark. "Current Problems in the Management of Marine Fisheries." *Science* 316 (2007): 1713–1716.

Beddington, J. R., and R. B. Rettig. *Approaches to the Regulation of Fishing Effort*. Food and Agricultural Organization (FAO) Fisheries Technical Paper no. 243. Rome: FAO, 1983.

Berkes, F., T. P. Hughes, R. S. Steneck, J. A. Wilson, D. R. Bellwood, B. Crona, C. Folke, et al. "Globalization, Roving Bandits, and Marine Resources." *Science* 311 (2006): 1557–1558.

Birnie, Patricia. *International Regulation of Whaling: From Conservation of Whaling to Conservation of Whales and Regulation of Whale Watching*. Vols. 1 and 2. New York: Oceana, 1985.

Bodansky, Daniel. *The Art and Craft of International Environmental Law*. Cambridge, MA: Harvard University Press, 2010.

Bonzon, Kate, Karly McIlwain, C. Kent Strauss, and Tonya Van Leuvan. *Catch Share Design Manual: A Guide for Managers and Fishermen*. Washington, DC: Environmental Defense Fund, 2010. http://www.edf.org/sites/default/files/catch-share-design-manual.pdf.

Brack, Duncan. *International Trade and the Montreal Protocol.* London: Royal Institute of International Affairs, 1996.

Branch, Trevor A. "The Influence of ITQs on Discarding and Fishing Behavior in Multispecies Fisheries." PhD diss., School of Aquatic and Fishery Sciences, University of Washington, Seattle, 2004.

Branch, Trevor A. "How Do Individual Transferable Quotas Affect Marine Ecosystems?" *Fish and Fisheries* 10 (2009): 39–57.

Branch, Trevor A. "A Review of Orange Roughy *Hoplostethus atlanticus* Fisheries, Estimation Methods, Biology, and Stock Structure." *African Journal of Marine Science* 23 (2001): 181–203.

Brown, Paul. "Soviet Union Illegally Killed Great Whales." *The Guardian*, 12 February 1994.

Buchanan, J. M. "An Economic Theory of Clubs." *Economica* 32 (1965): 1–14.

Buck, Eugene H. *Individual Transferable Quotas in Fishery Management.* Report for Congress no. 95-849. Washington, DC: Congressional Research Service, Library of Congress, 25 September 1995.

Buck, Susan. "No Tragedy of the Commons." *Environmental Ethics* 7 (Spring 1985): 49–61.

Burke, D. L., and G. L. Brander. "Canadian Experience with Individual Transferable Quotas." In *Use of Property Rights in Fisheries Management*, ed. R. Shotton, 151–160, Food and Agricultural Organization (FAO) Fisheries Technical Paper no. 401/1. Rome: FAO, 2000.

Burke, William T. "Fishing in the Bering Sea Donut: Straddling Stocks and the New International Law of Fisheries." *Ecology Law Quarterly* 16 (1989): 285–310.

Burtraw, Dallas. *Cost Savings, Market Performance, and Economic Benefits of the U.S. Acid Rain Program.* Discussion Paper no. 98-28-REV. Washington, DC: Resources for the Future, September 1998.

"Call Me Smiley." *New York Times Magazine* (13 March 1994): 14.

Campbell, Harry F., and R. B. Nicholl. "The Economics of the Japanese Tuna Fleet, 1979–80 to 1988–89." In *The Economics of Papua New Guinea's Tuna Fisheries*, ed. Harry F. Campbell and Anthony D. Owen, 39–52. Canberra: Australian Council for International Agricultural Research, 1994.

Campling, Liam, and Elizabeth Havice. "Mainstreaming Environment and Development at the WTO? The Peculiar Case of Fisheries Subsidies." Paper presented at the annual meeting of the International Studies Association, New Orleans, 17 February 2010.

Cantorna, Ana I. Sinde, Isabel Diéguez Castrillón, and Ana Gueimonde Canto. "Spain's Fisheries Sector: From the Birth of Modern Fishing through to the Decade of the Seventies." *Ocean Development & International Law* 38 (2007): 359–374.

Carpenter, Susan. *Special Corporations and the Bureaucracy: Why Japan Can't Reform.* Basingstoke, UK: Palgrave Macmillan, 2003.

Carr, Christopher J. "Recent Developments in Compliance and Enforcement for International Fisheries." *Ecology Law Quarterly* 24 (1997): 847–860.

CCAMLR (Commission for the Conservation of Antarctic Marine Living Resources). *Explanatory Memorandum on the Introduction Catch Documentation Scheme (CDS) for Toothfish*. N.d. http://www.ccamlr.org/pu/E/cds/p2.htm.

CCSBT (Commission for the Conservation of Southern Bluefin Tuna). *Comparison of CCSBT Catch Data with Japanese Auction Sales of Frozen SBT*. Working Paper no. CCSBT-ESC-0510/25. Deakin, Australia: CCSBT, 2005.

CCSBT (Commission for the Conservation of Southern Bluefin Tuna). *Report of the 15th Annual Meeting of the Commission (14–17 October 2008)*. Deakin, Australia: CCSBT, 2008. http://www.ccsbt.org/userfiles/file/docs_english/meetings/meeting_reports/ccsbt_15/report_of_CCSBT15.pdf

Chaffin, Joshua. "Cyber-theft Halts EU Emissions Trading." *Financial Times*, 19 January 2011.

"Charlemagne: A Commission Report-Card." *The Economist* (26 September 2009): 68.

Charles, Anthony T. "Use Rights and Responsible Fisheries: Limiting Access and Harvesting through Rights-Based Management." In *A Fishery Manager's Guidebook: Management Measures and Their Application*, ed. K. L. Cochrane, 131–157. Rome: Food and Agricultural Organization, 2002.

Charter, David. Too Many Boats Chasing Down Too Few Fish. *Sunday London Times*, 16 April 2009. http://www.timesonline.co.uk/tol/travel/news/article6101802.ece (accessed on 1 September 2011).

Chávez, Carlos, Nuria González, and Hugo Salgado. "ITQs under Illegal Fishing: An Application to the Red Shrimp Fishery in Chile." *Marine Policy* 32 (4) (2008): 570–579.

Chong, Florence. "Doha Round Will Fail 'Unless Members Agree.'" *The Australian*, 6 April 2011. http://www.theaustralian.com.au/business/doha-round-will-fail-unless-members-agree/story-e6frg8zx-1226034313883.

Chopin, Frank, Yoshihiro Inoue, Y. Matsushita, and Takafumi Arimoto. "Source of Accounted and Unaccounted Fishing Mortality." In *Solving Bycatch: Considerations for Today and Tomorrow*, ed. Alaska Sea Grant College Program, 41–47, Report no. 96-03. Fairbanks: University of Alaska, 1996.

Chu, Cindy. "Thirty Years Later: The Global Growth of ITQs and Their Influence on Stock Status in Marine Fisheries." *Fish and Fisheries* 10 (2) (2008): 217–230.

Clark, Colin W. "The Economics of Overexploitation." *Science* 181 (1973): 630–634.

Clark, Colin W. *Mathematical Bioeconomics: The Optimal Management of Renewable Resources*. 2nd ed. New York: Wiley-Interscience, 1990.

Clark, Colin W. *The Worldwide Crisis in Fisheries: Economic Models and Human Behavior*. Cambridge: Cambridge University Press, 2006.

Clucas, Ivor. *Fisheries Bycatch and Discards*. Food and Agricultural Organization (FAO) Fisheries Circular no. 928. Rome: FAO, 1997.

Coffey, Clare. "Fisheries Subsidies: Will the EU Turn Its Back on the 2002 Reforms?" World Wildlife Fund, May 2006. http://assets.panda.org/downloads/eu_fisheries_subsidies_2006.pdf.

Colson, David A. "Current Issues in Fishery Conservation and Management." *U.S. Department of State Dispatch* 6 (7) (1995). http://dosfan.lib.uic.edu/ERC/briefing/dispatch/1995/html/Dispatchv6no07.html.

Commission on Geosciences, Environment, and Resources. *An Assessment of Atlantic Bluefin Tuna*. Washington, DC: National Academies Press, 1994.

Committee on the Bering Sea Ecosystem, Polar Research Board, Commission on Geosciences, Environment, and Resources, and National Research Council. *The Bering Sea Ecosystem*. Washington, DC: National Academy Press, 1996.

Conference of the Parties to the Convention on Biological Diversity. *Report of the Second Meeting of the Conference of the Parties to the Convention on Biological Diversity*. UNEP/CBD/COP/2/19. Montreal: Convention on Biological Diversity Secretariat, 30 November 1995.

Constable, Andrew J., William K. de la Mare, David J. Agnew, Inigo Everson, and Denzil Miller. "Managing Fisheries to Conserve the Antarctic Marine Ecosystem: Practical Implementation of the Convention on the Conservation of Antarctic Marine Living Resources (CCAMLR)." *ICES Journal of Marine Science* 57 (2000): 778–791.

Convery, Frank J., and Luke Redmond. "Market and Price Developments in the European Union Emissions Trading Scheme." *Review of Environmental Economics and Policy* 1 (1) (1997): 88–111.

Cordell, John. "Carrying Capacity Analysis of Fixed-Territorial Fishing." *Ethnology* 17 (1) (1978): 1–24.

Corkett, Christopher. "The PEW Report on US Fishery Councils: A Critique from the Open Society." *Marine Policy* 29 (3) (2005): 247–253.

Cornes, Richard, Charles F. Mason, and Todd Sandler. "The Commons and the Optimal Number of Firms." *Quarterly Journal of Economics* 101 (3) (1986): 641–646.

Cornes, Richard, and Todd Sandler. *The Theory of Externalities, Public Goods, and Club Goods*. Cambridge: Cambridge University Press, 1996.

Costello, Christopher, Steven D. Gaines, and John Lynham. "Can Catch Shares Prevent Fisheries Collapse?" *Science* 321 (2008): 1678–1681.

Costello, Christopher J., and Daniel Kaffine. "Natural Resource Use with Limited-Tenure Property Rights." *Journal of Environmental Economics and Management* 55 (1) (2008): 20–36.

Cox, A. *Subsidies and Deep-Sea Fisheries Management: Policy Issues and Challenges*. Paris: Organization for Economic Cooperation and Development, n.d. http://www.oecd.org/dataoecd/10/27/24320313.pdf.

Cramton, Peter, and Suzi Kerr. "Tradeable Carbon Permit Auctions: How and Why to Auction Not Grandfather." *Energy Policy* 30 (4) (2002): 333–345.

Crothers, G. T., and Lindie Nelson. "High Seas Fisheries Governance: A Framework for the Future?" *Marine Resource Economics* 21 (2007): 341–353.

Cullis-Suzuki, Sarika, and Daniel Pauly. "Failing the High Sea: A Global Evaluation of Regional Fisheries Management Organizations." *Marine Policy* 34 (5) (2010): 1036–1042.

Cunningham, Steve, and Dominique Gréboval. *Managing Fishing Capacity: A Review of Policy and Technical Issues.* FAO Fisheries Technical Paper no. 409. Rome: FAO, 2001.

Dahou, Karim, Enda Tiers Monde, and M. Dème, with the collaboration of A. Dioum. "Support Policies to Senegalese Fisheries." In *Fisheries Subsidies and Maritime Resources Management: Lessons Learned from Studies in Argentina and Senegal*, ed. United Nations Environment Programme, 25–53. Geneva: United Nations Environment Program, 2002.

Daley, Beth. "Georges Bank Cod Stock on Decline." *Boston Globe*, 22 April 2003. http://www.eurocbc.org/page911.html.

Dauvergne, Peter, and Jane Lister. *Timber.* Cambridge: Polity Press, 2011.

Davie, Sarah, and Colm Lordan. "Examining Changes in Irish Fishing Practices in Response to the Cod Long-Term Plan." *ICES Journal of Marine Science* 68 (2011): 1638–1646.

Davies, R. W. D., S. J. Cripps, A. Nicksona, and G. Porter. "Defining and Estimating Global Marine Fisheries Bycatch." *Marine Policy* 33 (2009): 661–672.

Denmark Ministry of Food, Agriculture, and Fisheries. "Designing an ITQ Management: A Gain for Fishing Communities." 2010. http://www.fvm.dk/Files/Filer/Fiskeri/Danish_Design_of_ITQ.pdf.

DeSombre, Elizabeth R. *Domestic Sources of International Environmental Policy: Industry, Environmentalists, and U.S. Power.* Cambridge, MA: MIT Press, 2000.

DeSombre, Elizabeth R. "Fishing under Flags of Convenience: Using Market Power to Increase Participation in International Regulation." *Global Environmental Politics* 5 (4) (2005): 73–94.

DeSombre, Elizabeth R. *Flagging Standards: Globalization and Environmental, Safety, and Labor Standards at Sea.* Cambridge, MA: MIT Press, 2006.

DeSombre, Elizabeth R., and J. Samuel Barkin. *Fish.* Cambridge: Polity Press, 2011.

DeSombre, Elizabeth R., and J. Samuel Barkin. *The Turbot War: Canada, Spain, and Conflict over the North Atlantic Fishery.* Pew Case Study no. 226. Washington, DC: Institute for the Study of Diplomacy, 2000.

DeSombre, Elizabeth R., and J. Samuel Barkin. "Turtles and Trade: The WTO's Acceptance of Environmental Trade Restrictions." *Global Environmental Politics* 2 (1) (2002): 12–18.

Dower, John W. *Embracing Defeat: Japan in the Wake of WW II.* New York: Norton, 1999.

Dozier, Matt. "Gone Fishing." *National Marine Sanctuaries*, 12 April 2011. http://sanctuaries.noaa.gov/news/features/0411fishing.html.

"Draft Report of the Second Joint Meeting of Tuna Regional Fisheries Management Organizations (RFMOs)." 2nd Joint Tuna RFMOs Meeting, San Sebastian, Spain, 29 June–3 July 2009.

Ecotrust. *Catch-22: Conservation, Communities, and the Privatization of B.C. Fisheries: An Economic, Social, and Ecological Impact Study.* Portland, OR: Ecotrust, 2004. http://aquaticcommons.org/1686.

Eichenberg, Tim, and Mitchell Shapson. "The Promise of Johannesburg: Fisheries and the World Summit on Sustainable Development." *Golden Gate University Law Review* 34 (2004): 592–594.

El Serafy, Salah. "Green Accounting and Economic Policy." *Ecological Economics* 21 (3) (1997): 217–229.

Environmental Defense Fund. "Chilean National Benthic Resources Territorial Use Rights for Fishing Programme." Catch Share Design Center, 2011. http://apps.edf.org/catchshares/fishery.cfm?fishery_id=1802.

Environmental Defense Fund. "Maltese Kannizzati Fishery." Catch Share Design Center, 2011. http://apps.edf.org/catchshares/fishery.cfm?fishery_id=1271.

Environmental Defense Fund. "Swedish Coastal Territorial Use Rights for Fishing (TURF) System." Catch Share Design Center, 2011. http://apps.edf.org/catchshares/fishery.cfm?fishery_id=1327.

Environmental Defense Fund. "World Catch Share Database." 2011. http://www.edf.org/oceans/catch-share-design-center.

Essington, T. "Ecological Indicators Display Reduced Variation in North American Catch Share Fisheries." *Proceedings of the National Academy of Sciences of the United States* 107 (2) (2010): 754–759.

European Commission. "Bilateral Fisheries Partnership Agreements between the EC and Third Countries." 16 July 2009. http://ec.europa.eu/fisheries/cfp/external_relations/bilateral_agreements_en.htm (accessed on 29 July 2009).

European Commission. "Mauritania: Fisheries Partnership Agreement." Updated 17 February 2012. http://ec.europa.eu/fisheries/cfp/international/agreements/mauritania/index_en.htm.

European Court of Auditors. "Common Policy on Fisheries and the Sea." *Official Journal of the European Communities* (15 December 1992): 107–115.

Evans, Peter B., Harold K. Jacobson, and Robert D. Putnam, eds. *Double-Edged Diplomacy: International Bargaining and Domestic Politics.* Berkeley: University of California Press, 1993.

Eythórsson, Einar. "A Decade of ITQ-Management in Icelandic Fisheries: Consolidation without Consensus." *Marine Policy* 24 (2000): 483–492.

FAO (Food and Agriculture Organization). "Committee on Fisheries (COFI)—Fisheries and Aquaculture Department." 2011. http://www.fao.org/fishery/about/cofi/en.

FAO (Food and Agriculture Organization). "FAO Fisheries and Aquaculture Department—Mission." 2011. http://www.fao.org/fishery/about/organigram/en#Org-Mission/en.

FAO (Food and Agriculture Organization). *FAO Yearbook: Fishery and Aquaculture Statistics, 2007*. Rome: FAO, 2009.

FAO (Food and Agricultural Organization). "International Plan of Action for the Management of Fishing Capacity." 1999. http://www.fao.org/fishery/ipoa -capacity/legal-text/en.

FAO (Food and Agricultural Organization). "International Plan of Action to Prevent, Deter, and Eliminate Illegal, Unreported, and Unregulated Fishing." 2001. http://www.fao.org/docrep/003/y1224e/y1224e00.HTM.

FAO (Food and Agricultural Organization). *Managing Fishing Capacity of the World Tuna Fleet*. FAO Fisheries Circular no. 982, FIRM/C982(En) ISSN 0429-9329. Rome: FAO, 2003. http://www.fao.org/docrep/005/y4499e/y4499e00.htm.

FAO (Food and Agricultural Organization). "Marine Fisheries and the Law of the Sea: A Decade of Change." Special chapter (revised) of *State of Food and Agriculture 1992*, Marine Fisheries Circular No. 853, Rome: FAO, 1993.

FAO (Food and Agriculture Organization). *National Fisheries Sector Overview: Peru*. FID/CP/PER. Rome: FAO, May 2010). ftp://ftp.fao.org/FI/DOCUMENT/ fcp/en/FI_CP_PE.pdf.

FAO (Food and Agriculture Organization). "National Plans of Action—International Plan of Action for the Management of Fishing Capacity." 2011. http:// www.fao.org/fishery/ipoa-capacity/npoa/en.

FAO (Food and Agricultural Organization). *The State of the World Fisheries and Aquaculture*. Rome: FAO, 2008.

FAO (Food and Agricultural Organization). *The State of the World Fisheries and Aquaculture*. Rome: FAO, 2009.

FAO (Food and Agriculture Organization). "Statistics—Introduction." 2011. http://www.fao.org/fishery/statistics/en.

FAO (Food and Agriculture Organization), Fisheries and Aquaculture Department. "Fishery Statistics: Global Capture Production, 1950–2008." 2011. http:// www.fao.org/fishery/statistics/en.

FAO (Food and Agriculture Organization), Fisheries and Aquaculture Department. "Fishery Statistics: Spain Landings by Species, 1950–2009." 2011. http:// www.fao.org/fishery/statistics/en.

FAO (Food and Agricultural Organization), Marine Resources Service and Fisheries Department. *Review of the State of World Fishery Resources: Marine Fisheries*. FAO Fisheries Circular no. 920. FIRM/C920. Rome: FAO, 1997.

Fernandes, Deepali. *Running into Troubled Waters: The Fish Trade and Some Implications*. Evian Group Policy Brief. Geneva: Evian Group, November 2006. http://papers.ssrn.com/sol3/papers.cfm?abstract_id=1138271.

Ferris, J. S., and C. G. Plourde. "Labour Mobility, Seasonal Unemployment Insurance, and the Newfoundland Inshore Fishery." *Canadian Journal of Economics* 15 (3) (1982): 426–441.

Fisheries and Oceans Canada. "Kobe Process." 2011. http://www.dfo-mpo.gc.ca/ international/tuna-thon/Kobe-eng.htm.

Fisheries and Oceans Canada. "Recommended Best Practices for Regional Fisheries Management Organizations: A Model RFMO." 2009. http://www.dfo-mpo.gc.ca/international/documents/poster-eng.htm.

Fisheries and Oceans Canada. "What's Holding Back the Cod Recovery?" 2011. http://www.dfo-mpo.gc.ca/science/Publications/article/2006/01-11-2006-eng.htm.

"Fishing Redundancies Causes Union to Sue." *IceNews: News from the Nordics* (6 February 2008). http://www.icenews.is/index.php/2008/02/06/fishing-redundancies-causes-union-to-sue.

Fitzpatrick, J., and C. Newton. Assessment of the World's Fishing Fleet 1991–1997. Greenpeace, 5 May 1998. http://archive.greenpeace.org/oceans/global overfishing/assessmentfishingfleet.html.

Flaaten, O., and P. Wallis. *Government Financial Transfers to Fishing Industries in OECD Countries*. Paris: Organization for Economic Cooperation and Development, 2000.

Freeman, Jody, and Charles D. Kolstad, eds. *Moving to Markets: Lessons from Twenty Years of Experience*. New York: Oxford University Press, 2007.

French, Duncan. "Developing States and International Environmental Law: The Importance of Differentiated Responsibilities." *International and Comparative Law Quarterly* 49 (1) (2000): 35–60.

Gaines, Richard. "Whole Foods Agrees to Halt 'Red' Fish Sales." *Gloucester Daily Times*, 14 September 2010.

Gambles, Anna. "Free Trade and State Formation: The Political Economy of Fisheries Policy in Britain and the United Kingdom circa 1780–1950." *Journal of British Studies* 39 (3) (2000): 288–316.

Garcia, S. M. "The Precautionary Principle: Its Implications in Capture Fisheries Management." *Ocean and Coastal Management* 22 (2) (1994): 99–125.

Garcia, S. M., K. Cochrane, G. Van Santen, and F. Christy. Towards Sustainable Fisheries: A Strategy for FAO and the World Bank. *Ocean and Coastal Management* 42 (1999): 369–398.

Gates, John, Dan Holland, and Eyjolfur Gudmundsson. *Theory and Practice of Fishing Vessel Buy-Back Programs: Subsidies and Depletion of World Fisheries*. Washington, DC: World Wildlife Fund Endangered Seas Campaign, 1997.

Gauvin, John R., John M. Ward, and Edward E. Burgess. "Description and Evaluation of the Wreckfish (*Polyprion americanus*) Fishery under Individual Transferable Quotas." *Marine Resource Economics* 9 (1994): 99–118.

George, Jim. *Discourses of Global Politics: A Critical (Re)Introduction to International Relations*. Boulder, CO: Lynne Rienner, 1994.

Gianni, Matthew, and Walt Simpson. *Flags of Convenience, Transshipment, Re-Supply, and at-Sea Infrastructure in Relation to IUU Fishing*. Organization for Economic Cooperation and Development, Fisheries Committee, Directorate for Food, Agriculture, and Fisheries. AGR/FI/IUU. Rome: Food and Agricultural Organization, 2004.

Gibbs, M. "The Historical Development of Fisheries in New Zealand with Respect to Sustainable Development Principles." *Electronic Journal of Sustainable Development* 1 (2) (2008): 23–33.

Gillet, R., M. A. McCoy, and D. G. Itano. *Status of the United States Western Pacific Tuna Purse Seine Fleet and Factors Affecting Its Future.* Publication no. 02-01. Manoa: School of Ocean and Earth Science Technology, University of Hawaii, 2002.

Gjerde, Kristina. "Editor's Introduction: Moving from Words to Action." *International Journal of Marine and Coastal Law* 20 (3) (2005): 323–344.

Goddard, Stacie. "When Right Makes Might: How Prussia Overturned the European Balance of Power. *International Security* 33 (Winter 2008–2009): 110–142.

Goldsmith, Jack L., and Eric A. Posner. *Moral and Legal Rhetoric in International Relations: A Rational Choice Perspective.* John M. Olin Law and Economics Working Paper Series no. 108. Chicago: University of Chicago Law School, 2000.

Gómez-Lobo, A., J. Peña-Torres, and P. Barría. *ITQs in Chile: Measuring the Economic Benefits of Reform.* ILADES–Georgetown University Working Paper no. 179. Washington, DC: School of Economics and Business, Georgetown University, 2007.

Goode, Richard B. *Government Finance in Developing Countries.* Washington, DC: Brookings Institution Press, 1984.

Gordon, H. Scott. "The Economic Theory of a Common-Property Resource: The Fishery." *Journal of Political Economy* 62 (2) (1954): 124–142.

Goulder, Lawrence H., Ian W. H. Parry, Roberton C. Williams III, and Dallas Burtraw. "The Cost-Effectiveness of Alternative Instruments for Environmental Protection in a Second-Best Setting." *Journal of Public Economics* 7 (3) (1999): 329–360.

"Government to Promote Manta Port among Spanish Investors." *The Construction Gateway,* 28 December 2009. http://www.construpages.com/nl/noticia_nl.php?id_noticia=1191&language=en.

Grafton, R. Quentin. "Implications of Taxing Quota Value in an Individual Transferable Quota Fishery: Comment." *Marine Resource Economics* 11 (1996): 125–127.

Grafton, R. Quentin, Ragnar Arnason, Trond Bjørndal, David Campbell, Harry F. Campbell, Colin W. Clark, Robin Conner et al. "Incentive-Based Approaches to Sustainable Fisheries." *Canadian Journal of Fisheries and Aquatic Sciences* 63 (3) (2006): 699–710.

Graves, Theodore, and Dale Squires. "Lessons from Fisheries Buybacks." In *Fisheries Buybacks,* ed. Rita Curtis and Dale Squires, 15–53. Ames, IA: Blackwell, 2007.

Greenberg, Paul. *Four Fish: The Future of the Last Wild Food.* New York: Penguin, 2010.

Greenberg, Paul. "Tuna's End." *New York Times Magazine,* 27 June 2010.

Grotius, Hugo. *Mare Liberum*. 1618. Reprint. Oxford: Oxford University Press, 1916.

Groves, J. T., and D. Squires. "Requirements and Alternatives for the Limitation of Fishing Capacity in Tuna Purse-Seine Fleets." Paper Presented to the Food and Agricultural Organization Methodological Workshop on the Management of Tuna Fishing Capacity, La Jolla, CA, 8–12 May 2006.

Gruner, Peter. "The New Spanish Pirates." *Evening Standard*, 13 April 1995.

Haas, Peter M., Robert O. Keohane, and Marc A. Levy, eds. *Institutions for the Earth: Sources of Effective International Environmental Protection*. Cambridge, MA: MIT Press, 1993.

Hackett, Steven C. *Environmental and Natural Resources Economics*. 2nd ed. Armonk, NY: M. E. Sharpe, 2001.

Hamilton, Lawrence C., and Melissa J. Butler. "Outport Adaptations: Social Indicators through Newfoundland's Cod Crisis." *Human Ecology Review* 8 (2) (2001): 1–11.

Hanemann, W. Michael. "Valuing the Environment through Contingent Valuation." *Journal of Economic Perspectives* 8 (4) (Fall 1994): 19–43.

Hanich, Quentin, and Martin Tsamenyi. "Managing Fisheries and Corruption in the Pacific Islands Region." *Marine Policy* 33 (2) (2009): 386–392.

Hannesson, Rögnvaldur. "Effects of Liberalizing Trade in Fish, Fishing Services, and Investment in Fishing Vessels." Paper prepared for the Organization for Economic Cooperation and Development Committee on Fisheries, 87th Session, March 2001. http://www.oecd.org/dataoecd/1/11/1917250.pdf.

Hannesson, Rögnvaldur. "Growth Accounting in a Fishery." *Journal of Environmental Economics and Management* 53 (2007): 364–376.

Hannesson, Rögnvaldur. "Rights-Based Fishing on the High Seas: Is It Possible?" *Marine Policy* 35 (2011): 667–674.

Hardin, Garett. "The Tragedy of the Commons." *Science* 162 (1968): 1243–1248.

Hardin, Russell. *Collective Action*. Baltimore: Johns Hopkins University Press for Resources for the Future, 1982.

Havice, Elizabeth, and Liam Campling. "Shifting Tides in the Western and Central Pacific Ocean Tuna Fishery: The Political Economy of Regulation and Industry Responses." *Global Environmental Politics* 10 (1) (2010): 89–114.

Haward, Marcus, and Anthony Bergin. "The Political Economy of Japanese Distant Water Tuna Fisheries." *Marine Policy* 25 (2001): 91–101.

Hennessey, T. "Ludwig's Ratchet and the Collapse of New England Groundfish Stocks." *Coastal Management* 28 (2000): 187–213.

Hilborn, Ray. "Defining Success in Fisheries and Conflicts in Objectives." *Marine Policy* 31 (2) (2007): 153–158.

Hilborn, R., and C. J. Walters. *Quantitative Fish Stock Assessment: Choice, Dynamics, and Uncertainty*. New York: Chapman and Hall, 1992.

Holland, Dan, Eyjolfur Gudmundsson, and John Gates. "Do Fishing Vessel Buyback Programs Work? A Survey of the Evidence." *Marine Policy* 23 (1) (1998): 47–69.

Huppert, D., G. M. Ellis, and B. Noble. "Do Permit Prices Reflect the Discounted Value of Fishing? Evidence from Alaska's Commercial Salmon Fisheries." *Canadian Journal of Fisheries and Aquatic Sciences* 53 (4) (1996): 761–768.

IATTC (Inter-American Tropical Tuna Commission). "Proposal IATTC-82-J-1 Submitted by the European Union: Resolution for the Limitation of Fishing Capacity in Terms of Number of Active Longline Vessels." 82nd Meeting, La Jolla, CA, 4–8 July 2011. http://www.iattc.org/Meetings2011/Jun/PDFfiles/Proposals/IATTC-82-J-1-PROP-EUR-Limiting-longline-capacity.pdf.

ICCAT (International Commission for the Conservation of Atlantic Tunas). *2006 Report for the Biennial Period 2004–5*, Part II. Vol. 1. Madrid: ICCAT, 2006.

ICCAT (International Commission for the Conservation of Atlantic Tunas). *2007 Report for the Biennial Period 2006–7, PartI1*. Vol. 1. Madrid: ICCAT, 2007.

ICCAT (International Commission for the Conservation of Atlantic Tunas). *2008 Report for the Biennial Period 2006–7*. Part II. Vol. 1. Madrid: ICCAT, 2008.

ICCAT (International Commission for the Conservation of Atlantic Tunas). *2009 Report for the Biennial Period 2008–9*. Part I. Vol. 1. Madrid: ICCAT, 2009.

ICCAT (International Commission for the Conservation of Atlantic Tunas). *2010 Report for the Biennial Period 2008–9*. Part II. Vol. 1. Madrid: ICCAT, 2010.

ICCAT (International Commission for the Conservation of Atlantic Tunas). *2011 Report for the Biennial Period 2010*. Part I. Vol. 1. Madrid: ICCAT, 2011.

ICCAT (International Commission for the Conservation of Atlantic Tunas). *Recommendation by ICCAT Concerning the Bluefin Tuna Statistical Document Program*. Recommendation 92-1. Madrid: ICCAT, 1992.

Institute of Shipping Economics and Logistics. *ISL Shipping Statistics Yearbook 2003*. Bremen, Germany: ISL, 2004.

International Confederation of Free Trade Unions, Trade Union Advisory Committee to the OECD, ITC, and Greenpeace International. "More Troubled Waters: Fishing Pollutions and FOCs." 2002. Major Group Submission for the 2002 World Summit on Sustainable Development, Johannesburg.

International Pacific Halibut Commission. "About IPHC." N.d. http://www.iphc.washington.edu/about-iphc.html.

International Pacific Halibut Commission. "Commissioners." N.d. http://www.iphc.int/about-iphc/27.html.

IOTC (Indian Ocean Tuna Commission). "Approaches to Allocation Criteria in Other Tuna Regional Fishery Management Organizations." Prepared by the Secretariat, IOTC-2011-SS4-03[E], 2011. http://www.iotc.org/files/proceedings/2011/tcac/IOTC-2011-SS4-03%5BE%5D.pdf.

IOTC (Indian Ocean Tuna Commission). "Collection of Resolutions and Recommendations by the Indian Ocean Tuna Commission." Updated April 2010. http://www.iotc.org/English/resolutions.php.

IOTC (Indian Ocean Tuna Commission). *Report of the Thirteenth Session of the Scientific Committee.* Victoria, Australia: IOTC, 2011.

Iudicello, Suzanne, Michael L. Weber, and Robert Wieland. *Fish, Markets, and Fishermen: The Economics of Overfishing.* Washington, DC: Island Press, 1999.

Jamieson, Dale. "Scientific Uncertainty and the Political Process." *Annals of the AAPSS* 545 (1996): 35–43.

Japanese Ministry of Foreign Affairs. "Summary of the Joint Mission for Promoting Trade and Investment for Africa (the Central and West Mission)." 8 October 2008. http://www.mofa.go.jp/announce/announce/2008/10/1185334_1060.html.

"Japan's 1982 Fisheries Production Reaches Record High." *Marine Fisheries Review* 45 (7–9) (1983): 30.

Jeffries, B. *Catches by Other Countries.* Australian Fisheries Management Organization (AFMA) SBT-FAG/2000/20. Canberra: AFMA, 2000.

Jonsson, S. *The Development of the Icelandic Fishing Industry 1900–1940 and Its Regional Implications.* Reykjavik: Economic Development Institute, 1981.

Jorgensen, H., and C. Jensen. "Overcapacity, Subsidies, and Local Stability." In *Overcapacity, Overcapitalization, and Subsidies in European Fisheries,* ed. A. Hatcher and K. Robinson, 239–252. Portsmouth, UK: Centre for the Economics and Management of Aquatic Resources, University of Portsmouth, 1999.

Joseph, James. *Past Developments and Future Options for Managing Tuna Fishing Capacity, with Special Emphasis on Tuna Purse-Seine Fleets.* Rome: Food and Agricultural Organization, August 2005.

Joyner, Christopher C. "Compliance and Enforcement in New International Fisheries Law." *Temple International and Comparative Law Journal* 12 (Fall 1988): 271–299.

Kaczynski, Vladimir M. "Factory Motherships and Fish Carriers." *Journal of Contemporary Business* 10 (1) (1981): 59–74.

Kaczynski, Vladimir M., and David L. Fluharty. "European Policies in West Africa: Who Benefits from Fisheries Agreements?" *Marine Policy* 26 (2002): 75–93.

Kagoshima Prefecture Skipjack and Tuna Fisheries Cooperative Association. *Kagoshima Ken Katsuo Maguro Gyogyô Kyôdô Kumiai Sôritsu Gojûnenshu Shi* (Kagoshima Prefecture Skipjack and Tuna Fisheries Cooperative Association Fifty-Year History). Tokyo: Suisan Shinshio Sha, 2000.

Kahn, Ahmed S., U. Rashid Sumaila, Reg Watson, Gordon Munro, and Daniel Pauly. "The Nature and Magnitude of Global Non-fuel Fisheries Subsidies." In *Catching More Bait: A Bottom-Up Re-estimation of Global Fisheries Subsidies,* ed. Ussif Rashid Sumaila and Daniel Pauly, 5–37. Fisheries Centre Research Reports, vol. 14, no. 6. Vancouver: Fisheries Centre, University of British Columbia, 2006.

Kalland, Arne. *Fishing Villages in Tokugawa, Japan.* Honolulu: University of Hawaii Press, 1995.

Kariotis, Theodore C. "The Case of a Greek EEZ in the Aegean Sea." *Marine Policy* 14 (1) (1990): 3–14.

Kasahara, H. "Japanese Distant-Water Fisheries: A Review." *Fish Bulletin* 70 (1972): 227–282.

Keohane, Robert O. *After Hegemony: Cooperation and Discord in the World Political Economy*. Princeton, NJ: Princeton University Press, 1984.

Keohane, Robert O., Peter M. Haas, and Marc A. Levy. "The Effectiveness of International Environmental Institutions." In *Institutions for the Earth: Sources of Effective International Environmental Protection*, ed. Peter M. Haas, Robert O. Keohane, and Marc A. Levy, 3–24. Cambridge, MA: MIT Press, 1993.

Kingsley, M. C. "Food for Thought: ITQs and the Economics of High-Grading." *ICES Journal of Marine Science* 59 (2002): 649.

Kirkley, James E., Chris Reid, and Dale Squires. "Productivity Measurement in Fisheries with Undesirable Outputs: A Parametric, Non-stochastic Approach." Unpublished manuscript, 12 January 2010.

Knott, Jack H., and Gary J. Miller. *Reforming Bureaucracy: The Politics of Institutional Choice*. Englewood Cliffs, NJ: Prentice Hall, 1987.

Knudsen, Stale, and Hege Toje. "Post-Soviet Transformations in Russian and Ukrainian Black Sea Fisheries: Socio-economic Dynamics and Property Relations." *Southeast European and Black Sea Studies* 8 (1) (2008): 17–32.

Kock, Karl-Hermann, Keith Reid, John Croxall, and Stephen Nicol. "Fisheries in the Southern Ocean: An Ecosystem Approach." *Philosophical Transactions of the Royal Society, Biological Sciences* 362 (1488) (2007): 2333–2349.

Koring, Paul, and Kevin Cox. "Scots Support Seizure of Vessel." *Globe and Mail*, 11 March 1995.

Kuhn, Thomas. *The Structure of Scientific Revolutions*. Chicago: University of Chicago Press, 1962.

Kurlansky, Mark. *Cod: A Biography of the Fish That Changed the World*. New York: Walker, 1997.

Kurlansky, Mark. *The Last Fish Tale: The Fate of the Atlantic and Survival in Gloucester, America's Oldest Port and Most Original Town*. New York: Ballantine Books, 2008.

Lack, M., and G. Sant. "Patagonian Toothfish: Are Conservation and Trade Measures Working?" *TRAFFIC Bulletin* 19 (1) (2001): 1–18.

Laffont, Jean-Jacques, and Jean Tirole. "The Politics of Government Decision-Making: A Theory of Regulatory Capture." *Quarterly Journal of Economics* 106 (4) (1991): 1089–1127.

Lang, William L. "Beavers, Firs, Salmon, and Falling Water: Pacific Northwest Regionalism and the Environment." *Oregon Historical Quarterly* 104 (2) (Summer 2003): 150–165.

Layzer, Judith A. "Fish Stories: Science, Advocacy, and Policy Change in New England Fishery Management." *Policy Studies Journal* 3 (1) (2006): 59–80.

Layzer, Judith A. *Natural Experiments: Ecosystem-Based Management*. Cambridge, MA: MIT Press, 2008.

Le Gallic, Bertrand. "Why Is It Difficult for Governments to Move towards Using Market-Based Instruments in Fisheries?" Paper presented at the Fifteenth Annual European Association of Fisheries Economists Conference, Brest, 14–16 May 2003. http://www.oecd.org/dataoecd/50/27/15354941.pdf.

Levine, M. E. "Regulatory Capture, Public Interest, and the Public Agenda." *Journal of Law Economics and Organization* 6 (1990): 167–198.

Lipson, Charles. "International Cooperation in Economic and Security Affairs." *World Politics* 37 (1) (1984): 1–23.

Little, L. R., S. Kuikka, A. E. Putnam, F. Pantus, C. R. Davies, and B. D. Mapstone. "Information Flow among Fishing Vessels Modelled Using a Bayesian Network." *Environmental Modelling & Software* 19 (1) (2004): 27–34.

Lodahl, J. B., and G. Gordon. "Differences between Physical and Social Sciences in University Graduate Departments." *Research in Higher Education* 1 (1973): 191–213.

Lodahl, J. B., and G. Gordon. "Funding the Sciences in University Departments." *Educational Record* 54 (1973): 74–82.

Lodge, Michael W., David Anderson, Terje Løbach, Gordon Munro, Keith Sainsbury, and Anna Willock. *Recommended Best Practices for Regional Fisheries Management Organizations*. London: Royal Institute for International Affairs, 2007. http://www.chathamhouse.org/publications/papers/view/108473.

Ludwig, Donald, Ray Hilborn, and Carl Walters. "Uncertainty, Resource Exploitation, and Conservation: Lessons from History." *Science* 260 (1993): 17, 36.

Mace, Pamela. "A New Role for MSY in Single-Species and Ecosystem Approaches to Fisheries Stock Assessment and Management." *Fish and Fisheries* 2 (2001): 2–32.

Macinko, S., and D. W. Bromley. *Who Owns America's Fisheries?* Covelo, CA: Center for Resource Economics, 2002.

Marine Resources Assessment Group of the UK Department for International Development. *Fisheries and Access Agreements*. Policy Brief no. 6. London: MRAG, n.d.

Marshall, N. A. "Can Policy Perception Influence Social Resilience to Policy Change?" *Fisheries Research* 86 (2007): 216–227.

McCarthy, James E. *Clean Air Act: A Summary of the Act and Its Major Requirements*. Report for Congress Order Code RL30853. Washington, DC: Congressional Research Service, Library of Congress, updated 9 May 2005.

McGoodwin, James R. *Crisis in the World's Fisheries: People, Problems, and Politics*. Stanford, CA: Stanford University Press, 1990.

McMillan, John. "The Free-Rider Problem: A Survey." *Economic Record* 55 (2) (1979): 95–107.

Mikalsen, Knut, and Svein Jentoft. "From User Groups to Stakeholders? The Public Interest in Fisheries Management." *Marine Policy* 25 (4) (2001): 281–292.

Milazzo, Matteo. *Subsidies in World Fisheries: A Reexamination*. World Bank Technical Paper no. 406. Washington, DC: World Bank, 1998.

Miles, Edward, and William T. Burke. "Pressures on the United Nations Convention on the Law of the Sea of 1982 Arising from New Fisheries Conflicts: The Problem of Straddling Stocks." *Ocean Development and International Law* 20 (4) (1989): 343–357.

Ministry for Rural Affairs and the Environment (Malta), Department for International Development. *Fisheries Subsidies and the WTO Negotiations*. Policy Brief no. 9. Valletta, Malta: Ministry for Rural Affairs and the Environment, Department for International Development, 2009. http://www.mrag.co.uk/Documents/PolicyBrief9_Subsidies_insert_Apr09.pdf.

Mitchell, Bruce. "Politics, Fish, and International Resource Management: The British–Icelandic Cod War." *Geographical Review* 66 (2) (1976): 127–138.

Molyneaux, Paul. *Swimming in Circles: Aquaculture and the End of Wild Oceans*. New York: Thunder's Mouth Press, 2007.

Moravcsik, Andrew. "Why Is U.S. Human Rights Policy So Unilateralist?" In *Multilateralism and U.S. Foreign Policy: Ambivalent Engagement*, ed. Shepard Forman and Patrick Stewart, 435–476. Boulder, CO: Lynne Rienner, 2002.

Mullon, Christian, Pierre Fréon, and Philippe Cury. "The Dynamics of Collapse in World Fisheries." *Fish and Fisheries* 6 (2) (2005): 111–120.

Murias, Analia. "There Is No 'Possibility of Bluefin Tuna Collapse,' Says Association." *FIS Worldwide*, 8 October 2010. http://www.fis.com/fis/worldnews/worldnews.asp?monthyear=&day=8&id=38544&l=e&special=&ndb=1%20target=.

Myers, Ranson A., and Boris Worm. "Rapid Worldwide Depletion of Predatory Fish Communities." *Nature* 423 (2003): 280–283.

NAFO (Northwest Atlantic Fisheries Organization). *Meeting Proceedings of the General Council and Fisheries Commission for 2006/2007*. Dartmouth, Canada: NAFO, 2007.

NAFO (Northwest Atlantic Fisheries Organization). *Meeting Proceedings of the General Council and Fisheries Commission for 2009/2010*. Dartmouth, Canada: NAFO, 2010.

"Namibian Court Orders Seizure of Five Spanish Fishing Vessels." Radio Nacional de España, Madrid, *BBC Summary of World Broadcasts*, 12 April 1991. At Lexis/Nexis.

National Marine Fisheries Service, Alaska Regional Office. "BSAI Crab Rationalization FAQ." 2011. http://www.fakr.noaa.gov/sustainablefisheries/crab/rat/progfaq.htm#qsaifq.

National Oceanic and Atmospheric Administration, Fisheries Service. *Recreational Fisheries at NOAA: A Growing Legacy*. Washington, DC: National Marine Fisheries Service, 2011.

National Oceanic and Atmospheric Administration, US Department of Commerce. *Evaluating Bycatch: A National Approach to Standardized Bycatch Monitoring Programs*. National Oceanic and Atmospheric Administration (NOAA) Technical Memorandum NMFS-F/SPO-66. Washington, DC: NOAA, October 2004.

National Research Council. *Sharing the Fish: Toward a National Policy on Individual Fishing Quotas.* Washington, DC: National Academies Press, 1999.

NEAFC (North-East Atlantic Fisheries Commission). *Report of the 27th Annual Meeting of the North-East Atlantic Fisheries Commission, 10–14 November.* Vol. 1. London: NEAFC, 2008.

NEAFC (North-East Atlantic Fisheries Commission). *Report of the 28th Annual Meeting of the North-East Atlantic Fisheries Commission, 8–12 November.* London: NEAFC, 2010.

NEAFC (North-East Atlantic Fisheries Commission), Performance Review Panel. *Report of the North East Atlantic Fisheries Commission: Performance Review Panel Report of the NEAFC.* London: NEAFC, 6 November 2006. http://www. neafc.org/system/files/performance-review-final-edited.pdf.

Newell, R. G., J. N. Sanchirico, and S. Kerr. "Fishing Quota Markets." *Journal of Environmental Economics and Management* 49 (2005): 437–462.

Nielsen, Jesper Raakjaer, and Christoph Mathiesen. *Incentives for Compliance Behaviour: Lessons from Danish Fisheries.* Institute for Fisheries Management and Coastal Community Development Report no. 68. Exeter, UK: Institute for Fisheries Management and Coastal Community Development, 2001. http://www. ifm.dk/reports/68.pdf.

Nøstbakken, Linda, Olivier Thebaud, and Lars-Christian Sørensen. "Investment Behavior and Capacity Adjustment in Fisheries: A Survey of the Literature." *Marine Resource Economics* 26 (2) (2011): 95–118.

Noye, Jeremy, and Kame Mfodwo. "First Steps towards a Quota Allocation System in the Indian Ocean." *Marine Policy* 36 (2012): 882–894.

Oceana. *Global Fisheries Subsidies Regional Breakdown.* As derived from *Catching More Bait: A Bottom-Up Re-estimation of Global Fisheries Subsidies,* 2d version, ed. Ussif Rashid Sumaila and Daniel Pauly, Fisheries Centre Research Report, vol. 14, no. 6. Vancouver: Fisheries Centre, University of British Columbia, 5 June 2007. http://oceana.org/en/our-work/promote-responsible-fishing/ fishing-subsidies/learn-act/more-on-fisheries-subsidies.

OECD (Organization for Economic Cooperation and Development). *Draft Chapter 2—Framework for Measures against IUU Fisheries Activities.* 2004. AGR/FI/ IUU(2004)5/PROV. Paris: OECD, Fisheries Committee, Directorate for Food, Agriculture, and Fisheries.

OECD (Organization for Economic Cooperation and Development). *A General Procedure for Future Accessions.* C(2007)31/Final. Paris: OECD, 16 May 2007.

OECD (Organization for Economic Cooperation and Development). "Negotiation on a Shipbuilding Agreement." 2011. http://www.oecd.org/document/22/0,3 746,en_2649_34211_1823894_1_1_1_1,00.html.

OECD (Organization for Economic Cooperation and Development). *OECD Factbook 2010.* Paris: OECD, 2010.

OECD (Organization for Economic Cooperation and Development). *Review of Fisheries in OECD Countries 1997.* Paris: OECD, 1998.

OECD (Organization for Economic Cooperation and Development). *Review of Fisheries in OECD Countries 2009: Policies and Summary Statistics*. Paris: OECD, 2010.

OECD (Organization for Economic Cooperation and Development). *Review of Fisheries in OECD Countries: Country Statistics 2000–2002*. Paris: OECD, 2004.

OECD (Organization for Economic Cooperation and Development). "Shipbuilding Agreement—Overview." 2011. http://www.oecd.org/document/3/0,3343, en_2649_34211_1810179_1_1_1_1,00.html.

OECD (Organization for Economic Cooperation and Development). *Transition to Responsible Fisheries: Economic and Policy Implications*. Paris: OECD, 2000.

OECD (Organization for Economic Cooperation and Development). *Transition to Responsible Fisheries, Government Financial Transfers, and Resource Sustainability: Case Studies*. AGR/FI(2000)/10/FINAL. Paris: OECD, 2000.

Oelofsen, B., and A. Staby. "The Namibian Orange Roughy Fishery: Lessons Learned for Future Management." In *Deep Sea 2003: Conference on the Governance and Management of Deep-Sea Fisheries. Queenstown, New Zealand, 1–5 December 2003, Part 1: Conference Reports*, ed. R. Shotton, 555–559, Food and Agricultural Organization (FAO) Proceedings no. 3/1, Rome: FAO, 2005.

Olson, Mancur. *The Logic of Collective Action: Public Goods and the Theory of Groups*. Cambridge, MA: Harvard University Press, 1965.

Ostrom, Elinor. *Governing the Commons: The Evolution of Institutions for Collective Action*. Cambridge: Cambridge University Press, 1990.

Ostrom, Elinor, Roy Gardner, and James Walker. *Rules, Games, & Common-Pool Resources*. Ann Arbor: University of Michigan Press, 1994.

Pascoe, S. *Bycatch Management and the Economics of Discarding*. Food and Agricultural Organization (FAO) Technical Paper no. 370. Rome: FAO, 1997.

Pauly, Daniel, Reg Watson, and Jackie Adler. "Global Trends in World Fisheries: Impacts on Marine Ecosystems and Food Security." *Philosophical Transactions of the Royal Society, Biological Sciences* 360 (2005): 5–12.

Pautzke, C., and C. Oliver. *Development of the Individual Fishing Quota Program for Sablefish and Halibut Longline Fisheries Off Alaska*. Anchorage: North Pacific Fishery Management Council, 1997.

Pearson, Charles S. *Economics and the Global Environment*. Cambridge: Cambridge University Press, 2000.

Pencaval, John. "Labor Supply of Men: A Survey." In *The Handbook of Labor Economics*, ed. Orley Ashenfelter and Richard Layard, 3-102. New York: Elsevier, 1986.

Pfeffer, Jeffrey. "Barriers to the Advance of Organizational Science: Paradigm Development as a Dependent Variable." *Academy of Management Review* 18 (4) (1993): 599–620.

Phillips, Robert, R. Edward Freeman, and Andrew C. Wicks. "What Stakeholder Theory Is Not." *Business Ethics Quarterly* 13 (4) (2003): 479–502.

Polachek, Tom. "Politics and Independent Scientific Advice in RFMO Processes: A Case Study of Crossing Boundaries." *Marine Policy* 36 (2012): 132–141.

Polachek, Tom, and Campbell Davies. *Consideration of the Implications of Large Unreported Catches of Southern Bluefin Tuna for Assessments of Tropical Tuna, and the Need for Independent Verification of Catch and Effort Statistics.* Commonwealth Scientific and Industrial Research Organization (CSIRO) Marine and Atmospheric Research Paper no. 23. Clayton, Australia: CSIRO, 2008.

Poole, Erik. "Income Subsidies and Incentives to Overfish." In *Microbehavior and Macroresults: Proceedings of the Tenth Biennial Conference of the International Institute of Fisheries Economics and Trade Presentations,* 1–13. Corvalis, OR: Institute of Fisheries Economics and Trade, 2000.

Porch, Clay E. "The Sustainability of Western Atlantic Bluefin Tuna: A Warm-Blooded Fish in a Hot-Blooded Fishery." *Bulletin of Marine Science* 76 (2) (2005): 363–384.

Porter, Gareth. *Fisheries and the Environment: Subsidies and Overfishing, towards a Structured Discussion.* Geneva: United Nations Environment Program, 2001.

Porter, Gareth. *Fisheries Subsidies, Overfishing, and Trade.* Environment and Trade Report no. 16. Geneva: United Nations Environment Program, August 1998.

Powell, Robert. "Absolute and Relative Gains in International Relations Theory." *American Political Science Review* 85 (4) (1991): 1303–1320.

Putnam, Robert. "Diplomacy and Domestic Politics: The Logic of Two-Level Games." *International Organization* 42 (3) (Summer 1988): 427–460.

Rayfuse, Rosemary. *Regional Allocation Issues, or Zen and the Art of Pie-Cutting.* Faculty of Law Research Series, Paper no. 10. Sydney: University of New South Wales, 2007.

Reijinders, L. "Subsidies and the Environment." In *Producer Subsidies,* ed. R. Gerritse, 111–121. London: Pinter, 1990.

"Report of the Tuna RFMOs Chairs' Meeting." San Francisco, 5–6 February 2008. http://www.tuna-org.org/Documents/RFMO_CHAIRS_FEB%205-6_FRISCO _Phils.pdf.

Risse, Thomas. "International Norms and Domestic Change: Arguing and Communicative Behavior in the Human Rights Area." *Politics & Society* 27 (4) (1999): 529–559.

Roberts, Callum. *The Unnatural History of the Sea.* Washington, DC: Island Press, 2007.

Rose, Alex. *Who Killed the Grand Banks? The Untold Story behind the Decimation of One of the World's Greatest Natural Resources.* Mississauga, Canada: Wiley, 2008.

Rose, R. *Efficiency of Individual Transferable Quotas in Fisheries Management.* Australian Bureau of Agriculture and Resource Economics (ABARE) Report to the Fisheries Resources Research Fund. Canberra: ABARE, September 2002.

Ruckleshaus, Mary, Terrie Klinger, Nancy Knowlton, and Douglas P. DeMaster. "Marine Ecosystem–Based Management in Practice: Scientific and Governance Challenges." *BioScience* 58 (1) (2008): 53–63.

Russin, Krystle. "NOAA Eliminates Competitive Fish Harvesting." *Heartland Institute Environment & Climate News* (October 2010): 4.

Salz, P. *The European Atlantic Fisheries*. The Hague: Agricultural Economics Research Institute, 1991.

Sano, Hiroya. *Are Private Initiatives a Possible Way Forward? Actions Taken by Private Stakeholders to Eliminate IUU Fishing Activities*. AGR/FI/IUU(2004)13. Paris: Organization for Economic Cooperation and Development, Fisheries Committee, Directorate for Food, Agriculture, and Fisheries, 8 April 2004.

Schiffman, Howard S. *Marine Conservation Agreements: The Law and Policy of Reservations and Vetoes*. Leiden: Nijhoff, 2008.

Schorr, David K. *Healthy Fisheries, Sustainable Trade: Crafting New Rules on Fishing Subsidies in the World Trade Organization*. World Wildlife Fund (WWF) Position Paper and Technical Resource. Washington, DC: WWF, June 2004.

Schrank, William E. *Introducing Fisheries Subsidies*. Food and Agricultural Organization (FAO) Fisheries Technical Paper no. 437. Rome: FAO, 2003.

Scott, Anthony. "Conceptual Origins of Rights-Based Fishing." In *Rights Based Fishing: Proceedings of the NATO Advanced Research Workshop on Scientific Foundations for Rights-Based Fishing, 27 June–1 July 1988, in Reykjavik, Iceland*, ed. Philip A. Neher, Ragnar Arnason, and Nina Mollett, 11–38. Dordrecht: Kluwer Academic, 1989.

Scott, Anthony. "The Fishery: The Objectives of Sole Ownership." *Journal of Political Economy* 63 (2) (1955): 116–124.

Shahidul Islam, M., and Masaru Tanaka. "Impacts of Pollution on Coastal and Marine Ecosystems Including Coastal and Marine Fisheries and Approach for Management: A Review and Synthesis." *Marine Pollution Bulletin* 48 (2004): 624–649.

Sibert, John, John Hampton, Pierre Kleiber, and Mark Maunder. "Biomass, Size, and Trophic Status of Top Predators in the Pacific Ocean." Science 314 (2006): 1773–1776.

Simon, Julian. *The Ultimate Resource*. Princeton, NJ: Princeton University Press, 1998.

Sinclair, M., R. N. O'Boyle, D. L. Burke, and F. G. Peacock. "Groundfish Management in Transition within the Scotia-Fundy Area of Canada." *ICES Journal of Marine Science* 56 (1999): 1014–1023.

Sissenwine, M. P., and P. M. Mace. "ITQs in New Zealand: The Era of Fixed Quota in Perpetuity." *Fish Bulletin* 90 (1992): 147–160.

Smith, Robert W. *Exclusive Economic Zone Claims: An Analysis and Primary Documents*. Dordrecht, Netherlands: Nijhoff, 1986.

Smith, Zachary A. *The Environmental Policy Paradox*. 2nd ed. Englewood Cliffs, NJ: Prentice Hall, 1995.

Spagnolo, Massimo, and Evelina Sabatella. "Driftnets Buy Back Program: A Case of Institutional Failure." In *Fisheries Buybacks*, ed. Rita Curtis and Dale Squires, 145–156. Ames, IA: Blackwell, 2007.

Spenser, Thomas. "Environmental Groups Plan Lawsuit on Petition to Protect Species." *Birmingham News*, 22 April 2011. http://www.biologicaldiversity.org/news/center/articles/2011/birmingham-news-04-22-2011.html.

Steele, Scott. "A Haul of Hard Cash." *Maclean's*, 4 August 1997.

Stiglitz, Joseph. "A Neoclassical Analysis of the Economics of Natural Resources." In *Scarcity and Growth Reconsidered*, ed. V. K. Smith, 36–66. Baltimore: Johns Hopkins University Press, 1979.

Stiglitz, Joseph E., Amartya Sen, and Jean-Paul Fitoussi. *Mis-measuring Our Lives: Why GDP Doesn't Add Up*. New York: New Press, 2010.

Stone, Christopher D. "Common but Differentiated Responsibilities in International Law." *American Journal of International Law* 98 (2) (2004): 276–301.

Streck, Charlotte. "The Global Environment Facility—a Role Model for International Governance?" *Global Environmental Politics* 1 (2) (2001): 71–94.

Sumaila, Ussif Rashid, Ahmed Kahn, Louise The, Reg Watson, Peter Tyedmers, and Daniel Pauly. "Subsidies to High Seas Bottom Trawl Fleets and the Sustainability of Deep Sea Benthic Stock." In *Catching More Bait: A Bottom-Up Re-estimation of Global Fisheries Subsidies*, ed. Ussif Rashid Sumaila and Daniel Pauly, 49–53. Fisheries Centre Research Reports, vol. 14, no. 6. Vancouver: Fisheries Centre, University of British Columbia, 2006.

Sumaila, Ussif Rashid, and Daniel Pauly, eds. *Catching More Bait: A Bottom-Up Re-estimation of Global Fisheries Subsidies*. Fisheries Centre Research Reports, vol. 14, no. 6. Vancouver: Fisheries Centre, University of British Columbia, 2006.

Sumaila, Ussif Rashid, Louise The, Reg Watson, Peter Tyedmers, and Daniel Pauly. "Fuel Subsidies to Global Fisheries: Magnitude and Impacts on Resource Sustainability." *Catching More Bait: A Bottom-Up Re-estimation of Global Fisheries Subsidies*, ed. Ussif Rashid Sumaila and Daniel Pauly, 38–48. Fisheries Centre Research Reports, vol. 14, no. 6. Vancouver: Fisheries Centre, University of British Columbia, 2006.

Swan, Judith. *Fishing Vessels Operating under Open Registers and the Exercise of Flag State Responsibilities—Information and Options*. Food and Agricultural Organization (FAO) Fisheries Circular no. 980. FITT/C980. Rome: FAO, 2002.

Symes, David. *The Integration of Fisheries Management and Marine Wildlife Conservation*. Joint Nature Conservation Committee (JNCC) Report no. 287. Peterborough: JNCC, 1998. http://jncc.defra.gov.uk/pdf/RPT287.pdf.

Teece, David R. "Global Overfishing and the Spanish–Canadian Turbot War: Can International Law Protect the High-Seas Environment?" *Journal of International Environmental Law and Policy* 89 (1997): 89–125.

Tietenberg, Thomas H. *Emissions Trading: Principles and Practice*. 2nd ed. Washington, DC: Resources for the Future, 2006.

Tietenberg, Thomas H. *Environmental and Natural Resource Economics*. 5th ed. Reading, MA: Addison-Wesley, 2000.

Turris, B. R. "A Comparison of British Columbia's ITQ Fisheries for Groundfish Trawl and Sablefish: Similar Results from Programmes with Differing Objectives, Designs, and Processes." In *Use of Property Rights in Fisheries Management*, ed. R. Shotton, 254–261. Food and Agricultural Organization (FAO) Fisheries Technical Paper no. 404/1. Fremantle, Australia: FAO, 2000.

UNEP (United Nations Environment Programme). *Analyzing the Resource Impact of Fisheries Subsidies: A Matrix Approach*. Geneva: UNEP, 2004.

UNEP (United Nations Environment Programme). *Ecosystem Management Programme: A New Approach to Sustainability*. Nairobi: UNEP, 2009.

UNEP (United Nations Environment Programme). *Incorporating Resource Impacts into Fisheries Subsidies Disciplines: Issues and Options, a Discussion Paper*. UNEP/ETB/2004/10. Nairobi: UNEP, 2004. http://www.unep.ch/etu/Fisheries%20Meeting/IncorResImpFishSubs.pdf.

UNEP (United Nations Environment Programme). *Towards a Green Economy: Pathways to Sustainable Development and Poverty Eradication*. Nairobi: UNEP, 2011.

UNEP (United Nations Environment Programme), Division of Technology, Industry, and Economics. "Subsidies in Argentine Fisheries." UNEP Fisheries Workshop, Geneva, 12 February 2001. http://www.unep.ch/etb/events/events2001/fishery/fisheryArgentina.pdf.

United Nations Industrial Development Organization. *Public Goods for Economic Development*. Vienna: United Nations Industrial Development Organization, 2008.

United Nations Statistical Division. *Integrated Environmental and Economic Accounting: Handbook of National Accounting*. Geneva: United Nations, 1993.

US Department of the Interior, Fish and Wildlife Service, and US Department of Commerce, US Census Bureau. *2006 National Survey of Fishing, Hunting, and Wildlife-Associated Recreation*. Washington, DC: US Fish and Wildlife Service, 2006.

US Environmental Protection Agency. "Acid Rain Program Benefits Exceed Expectations." N.d. http://www.epa.gov/capandtrade/documents/benefits.pdf.

US General Accounting Office. *International Environment: International Agreements Are Not Well-Monitored*. GAO-RCED-92–43. Washington, DC: US General Accounting Office, 1992.

van Bochove, Christiaan. "The 'Golden Mountain': An Economic Analysis of Holland's Early Modern Herring Fisheries." In *Beyond the Catch: Fisheries of the North Atlantic, the North Sea, and the Baltic, 900–1850*, ed. Louis Sicking and Darlene Abreu-Ferreira, 209–242. Leiden: Brill, 2008.

Wallis, P., and O. Flaaten. *Fisheries Management Cost: Concepts and Studies*. Rome: Food and Agricultural Organization, 2000.

Walsh, Mary Williams. "Canada Slashes Cod Quota in Fishing Crisis." *Los Angeles Times*, 25 February 1992. http://articles.latimes.com/1992-02-25/news/mn-2657_1_cod-population.

Walters, Carl, Villy Christensen, Steven Martell, and James Kitchell. "Possible Ecosystem Impacts of Applying MSY Policies from Single-Species Assessment." *ICES Journal of Marine Science* 62 (3) (2005): 558–568.

Walters, Carl J., and Steven J. D. Martell. *Fisheries Ecology and Management.* Princeton, NJ: Princeton University Press, 2004.

WCPFC (Western and Central Pacific Fisheries Commission). "Commission for the Conservation and Management of Highly-Migratory Fish Stocks in the Western and Central Pacific Ocean, Fifth Regular Session. Busan, Korea, 8–12 December 2008." 2009. http://www.wcpfc.int/node/1892.

Webster, D. G. *Adaptive Governance: The Dynamics of Atlantic Fisheries Management.* Cambridge, MA: MIT Press, 2009.

Weiss, Carol H., and Allen H. Barton, eds. *Making Bureaucracies Work.* Beverly Hills, CA: Sage, 1979.

Wilen, J. E., and E. J. Richardson. "Rent Generation in the Alaskan Pollock Conservation Cooperative." In *Case Studies in Fisheries Self-Governance,* ed. Ralph Edwin Townsend, Ross Shotton, and Hirotsugu Uchida, 361–368. FAO Fisheries Technical Paper no. 504. Rome: FAO, 2008.

Willock, A., and M. Lack. *Follow the Leader: Learning from Experience and Best Practice in Regional Fishery Management Organizations.* Washington, DC: World Wildlife Fund International and TRAFFIC International, 2006.

Wingard, John D. "Community Transferable Quotas: Internalizing Externalities and Minimizing Social Impacts of Fisheries Management." *Human Organization* 59 (2000): 48–57.

World Bank. *What Is PROFISH? The World Bank's Program on Fisheries.* Washington, DC: World Bank, September 2009. http://siteresources.worldbank.org/EXTARD/Resources/336681-1224775570533/3WhatsPROFISH.pdf.

World Bank, Agricultural and Rural Development Department. *Saving Fish and Fishers: Toward Sustainable and Equitable Governance of the Global Fishing Sector.* Report no. 29090-GLB. Washington, DC: World Bank, May 2004.

Worm, Boris, Edward B. Barbier, Nicola Beaumont, J. Emmett Duffy, Carl Folke, Benjamin S. Halpern, Jeremy B. C. Jackson, et al. "Impacts of Biodiversity Loss on Ocean Ecosystem Services." *Science* 314 (2006): 787–790.

WTO (World Trade Organization). "Doho Declaration Explained." 2011. http://www.wto.org/english/tratop_e/dda_e/dohaexplained_e.htm.

WTO (World Trade Organization). *The Environmental Benefits of Removing Trade Restrictions and Distortions: The Fisheries Sector.* Note by the Secretariat. WT/CTE/W/167. Geneva: WTO, 16 October 2000.

WTO (World Trade Organization). "Negotiations on Fisheries Subsidies." 2001. http://www.wto.org/english/tratop_e/rulesneg_e/fish_e/fish_e.htm.

WTO (World Trade Organization), Negotiating Group on Rules. *Draft Consolidated Chair Texts of the AD and SCM Agreements.* TN/RL/W/213. Geneva: WTO, 30 November 2007.

WWF (World Wildlife Fund). *Hard Facts, Hidden Problems: A Review of Current Data on Fisheries Subsidies.* Washington, DC: WWF, 2001.

WWF (World Wildlife Fund). *Small Boats, Big Problems.* Gland, Switzerland: WWF, 2008. http://www.illegal-fishing.info/uploads/wwfsmallboatsbigproblems1 .pdf.

Young, Margaret. "Fragmentation or Interaction: The WTO, Fisheries Subsidies, and International Law." *World Trade Review* 8 (2009): 477–515.

Zarrilli, Luca. "Iceland and the Crisis: Territory, Europe, Identity." *Revista Română de Geografie Politică* 8 (1) (2011): 5–15.

Zhuang, Juzhong, Zhihong Liang, Tun Lin, and Franklin De Guzman. *Theory and Practice in the Choice of Social Discount Rate for Cost–Benefit Analysis: A Survey.* European Report on Development Working Paper Series no. 94. Manilla: Asian Development Bank, May 2007.

Index